Alexis L. Boylan, Anna Mae Duane, Michael Gill, and Barbara Gurr
Furious Feminisms: Alternate Routes on *Mad Max: Fury Road*

Ian G. R. Shaw and Marv Waterstone
Wageless Life: A Manifesto for a Future beyond Capitalism

Claudia Milian
LatinX

Aaron Jaffe
Spoiler Alert: A Critical Guide

Don Ihde
Medical Technics

Jonathan Beecher Field
Town Hall Meetings and the Death of Deliberation

Jennifer Gabrys
How to Do Things with Sensors

Naa Oyo A. Kwate
Burgers in Blackface: Anti-Black Restaurants Then and Now

Arne De Boever
Against Aesthetic Exceptionalism

Steve Mentz
Break Up the Anthropocene

John Protevi
Edges of the State

Matthew J. Wolf-Meyer
Theory for the World to Come: Speculative Fiction and Apocalyptic Anthropology

Nicholas Tampio
Learning versus the Common Core

Kathryn Yusoff
A Billion Black Anthropocenes or None

Kenneth J. Saltman
The Swindle of Innovative Educational Finance

Ginger Nolan
The Neocolonialism of the Global Village

Joanna Zylinska
The End of Man: A Feminist Counterapocalypse

(Continued on page 114)

Does the Earth Care?

Forerunners: Ideas First

Short books of thought-in-process scholarship, where intense analysis, questioning, and speculation take the lead

FROM THE UNIVERSITY OF MINNESOTA PRESS

Mick Smith and Jason Young
Does the Earth Care? Indifference, Providence, and Provisional Ecology

Caterina Albano
Out of Breath: Vulnerability of Air in Contemporary Art

Gregg Lambert
The World Is Gone: Philosophy in Light of the Pandemic

Grant Farred
Only a Black Athlete Can Save Us Now

Anna Watkins Fisher
Safety Orange

Heather Warren-Crow and Andrea Jonsson
Young-Girls in Echoland: #Theorizing Tiqqun

Joshua Schuster and Derek Woods
Calamity Theory: Three Critiques of Existential Risk

Daniel Bertrand Monk and Andrew Herscher
The Global Shelter Imaginary: IKEA Humanitarianism and Rightless Relief

Catherine Liu
Virtue Hoarders: The Case against the Professional Managerial Class

Christopher Schaberg
Grounded: Perpetual Flight . . . and Then the Pandemic

Marquis Bey
The Problem of the Negro as a Problem for Gender

Cristina Beltrán
Cruelty as Citizenship: How Migrant Suffering Sustains White Democracy

Hil Malatino
Trans Care

Sarah Juliet Lauro
Kill the Overseer! The Gamification of Slave Resistance

Does the Earth Care?

Indifference, Providence, and Provisional Ecology

Mick Smith and Jason Young

University of Minnesota Press

MINNEAPOLIS

LONDON

ISBN 978-1-5179-1320-5 (PB)
ISBN 978-1-4529-6706-6 (Ebook)
ISBN 978-1-4529-6847-6 (Manifold)

Published by the University of Minnesota Press, 2022
111 Third Avenue South, Suite 290
Minneapolis, MN 55401-2520
http://www.upress.umn.edu

Available as a Manifold edition at manifold.umn.edu

Dedicated in loving memory of Ruby and Jim Davidson,
and to a dear friend, Sandra Spence

Contents

Preface

TO SPEAK OF EARTHLY PROVIDENCE would seem anachronistic and dubious in a world of devastating pandemics, mass extinction, and destabilizing climate change. Nevertheless, rethinking the legacy of that providential imaginary which prospered in pre- and early modern Europe might still provide certain affordances for understanding and expressing responses to such issues. For all its faults, *and there are many,* this providential imaginary found, in the earth's manifestations, signs of a world infused with care, albeit a care purportedly originating in God and directed primarily, or only, toward (certain) humans.

Tracing how this providential imaginary evolved into a succeeding (and globally dominant) "progressive" imaginary, we can witness how it was transformed in ways that still allowed its adherents to consider themselves "elect" or "favored" while evacuating an increasingly objectivized earth of all but the most superficial relations to care. That is to say, the imaginary of progress, including scientific progress, offered no respite from, indeed helped generate, a "systemic worldly indifference" such that referring to the earth as "caring" can only appear ridiculously and unjustifiably anthropomorphic or indicative of an antiquated wishful thinking. This absence of earthly solicitude should, we suggest, be considered a symptom and not just a consequence of this predominant modern mode of materio-semiotic engagement with the earth, now exemplified in global capitalism. Here the earth is systematically reduced to a largely inanimate resource in the service of the very projects that instigate the climatological changes now undermining this same world. In other words, the "successful" globalization of the progressive imaginary is revealed by climate change to be self-abnegating; its explosive teleology is now shown to lead only to an ignominious end of this world and the extinction of so many of its diverse inhabitants. Earth, though, will persist.

At issue, then, is how we might differently address, and also be materially and semiotically addressed by, the earth and how the earth *matters* beyond (human) signification. Certainly there is no single concept or metaphor adequate for engendering or conveying an alternative earthly imaginary. Earth is not a spaceship, a system, or a goddess, nor is it just another rock circling the sun, a superorganism, our birthright, and so on. This inadequacy occurs because of the expressive limits of any language and because the earth subtends and transcends the historically inflected worlds we inhabit, those worlds that matter and mean something to us. Earth, we might say, resists globalization, including conceptual globalization, and is both refractory and exceeds all attempts to encompass or enframe "it," even technologically. After all, not everything can be integrated, synthesized, or reconciled (Baudrillard 2007, 67). This situation might, perhaps, be approached through Heidegger's later philosophy, where *earth* and *world* are, of course, terms of art, but it is also, very importantly, something that can be *experienced* as manifestations of earthly indifference, care, and so on. Such experiences are something that philosophically inclined "nature" writers like Robert Macfarlane and Richard Jefferies explicitly address, which is why we begin and end with such accounts.

Despite these expressive difficulties, any ecological alternatives to the providential and progressive imaginaries that have so dramatically shaped our world would obviously need to offer some way to *provisionally* reconceptualize their legacies and to explicate very different, less destructive understandings of our earthly existence. Such a "provisional ecology," albeit tentatively expressed and necessarily incomplete, must explicitly acknowledge how and what the earth provides.

This is where discussions around James Lovelock's (1979) development of the Gaia hypothesis (originally developed with Lynn Margulis), including Bruno Latour's (2017a, 2017b) recent and extensive interventions, might prove generative. Interestingly, although Lovelock's work now provides key systemic insights used in contemporary climatological work, scientific critics often interpreted the planetary homeostasis of Gaia's earliest iterations as impossibly *providential* and doubtfully anthropomorphic. However, in his more explicitly scientific moments, and in Latour's (2017b, 69) subsequent claim that Gaia is "totally non-providential," the earth is often portrayed as entirely indifferent to humanity's clima(c)tic fate. To the extent that Lovelock's texts actually travel back and forth between these poles, Gaia might serve as an opening on the tensions between earthly indifference and care, and thence on arguments we make here for understanding the earth in terms of its expressive composition of *provisional* (temporary, contingent, and yet creatively providing) ecologies. The claim here is that a provisional rather than providential understanding of earthly ecology might prove a more appropriate counter to forms of cold "realism" and/or alienated despair in the face of impending ecological disaster.

Prologue: Earthly Indifference

> I have often sensed the indifference of matter in the mountains and found it exhilarating. But the black ice exhibited another order of withdrawnness, one so extreme as to induce nausea. (Macfarlane 2019, 381)

IN AN AGE OF CLIMATE CHANGE, even nature writers, such as Macfarlane, sometimes encounter a profound sense of nature's indifference. The "black ice" he describes is the ancient progeny of a Greenland glacier calving prematurely due to climate change. One hundred thousand years in the making, and weighing "hundreds of thousands of tons," the blue-white front of the glacier that had crashed down into ocean waters suddenly and shockingly resurfaces as a vast, black, metallic mass, a huge displaced alien "something," "that has come from so deep down in time that it has lost all color" (Macfarlane 2019, 377). This preternaturally dark mountain of ice surges up, displacing and shedding its watery contours into a world now entirely altered: a revelation of immense, yet previously unacknowledged, material powers. And yet, even as this manifestation of sublime alienation subsides, it will begin to melt away, leaving no identifiable trace, except, perhaps, an almost imperceptible addition to rising sea levels. What, if anything, can this apparition mean? Who could say? A consequence of "anthropogenic" climate change, a sign of the times, perhaps even of the end of "our" time here on earth? Even the most objective climate science might be tempted to read it as a portent.

How different this seems from the ice grains, caught in their dazzling and entrancing play with bright sunlight, encroaching and sparkling on the windowpane. How different from the foot-deep snow that, for today at least, before it, too, liquefies with spring's rising temperatures, blankets the land. Yet both poles, the apparently homely and the experientially *unheimlich*—the uncanny and alienating—are created through more, or less, mediated encounters with ostensibly the same life-giving substance. Part of the issue here, then, is how to say anything intelligible about the mutability and multiplicitous agency of material forms; their vastly different experiential effects; the atmospheres they create and destroy; and even their resistance to being appropriated, colonized, ordered, explained, or assimilated, by human language itself. For where the window's ice crystals joyfully distract, the iceberg's appearance momentarily threatens to overwhelm and dissolve the anthropocentric tales modernity tells of its power and progress, words that, for centuries now, have striven to make the world nothing but an object of our possession.

> Here was a region where matter drove language aside. Ice left language beached. The object refused its profile. Ice would not mean, nor would rock or light, and so this was a weird realm, in the old strong sense of weird—a terrain that could not be communicated in human terms or forms. (Macfarlane 2019, 381)

Notice that Macfarlane does not, in any way, dismiss this manifestation as meaningless. On the contrary, the iceberg is hugely significant, but its significance emerges in its sudden disruption of the sense of our own importance, our centrality in the stories we tell, the words we constantly employ to make the world meaningful to us. It sunders the already fraying warp and weft of language that had served to weave together a modern world we had assumed was ours to shape and master. It is these narratives, and not the berg itself, that now begin to appear (as tales told by fools) all "sound and fury, signifying nothing" (Shakespeare 1953, 553).

From a distance (whether geographic, textual, theoretical, or temporal), the calving glacier might certainly be taken to exemplify anthropo*genic* climate change, yet experientially, being-there at that moment, the berg's world-changing reemergence is an *event* (in a Heideggerian [1989][1] sense) that exposes the profound limits to human understanding and the facile assumption that all that really *matters* lies within "our" purview. The berg shatters any illusion that all meaning can be contained and expressed in (modern) human language and subverts the assumption that it is what "we" do and say that bestows meaning on a pliant earth. This event also challenges the dominant modern presumption that nature, in its indifferent objectivity, can have no possible message for us.

If, for Heidegger (1977a, 213), "language is the house of Being," then we might say that the berg refuses to be domesticated. The berg's appearance is a shock to the system and to every frame of reference. Its appearance is prodigious (Burns 2002), a *freak of nature*, and hence simultaneously natural and unnatural, an expression of an age long past and of eons yet to unfold without us. It is *ab*normal in refusing our categorizations, in exhibiting its incompatibility with our mundane expectations, a momentous revelation of anthropic precarity. The iceberg weirds us out through its forceful revelation of the material and temporal immensity of nature; it reveals, almost instantaneously, the manifest absurdity of every human life and the inevitability of humanity's eventual extinction.

Later, of course, as witnesses after the event, we are drawn to relate (ourselves to) this revelation. For if the berg insists on being experienced coldly, nihilistically, even despairingly, then to voice this is still to recuperate it in its relation to human being, if only in

1. In appropriating aspects of Heidegger's philosophy, we certainly recognize concerns regarding his political association with national socialism in the 1930s and his personal anti-Semitism. The degree to which these inform, or are expressed at all in, his philosophical work continues to be a matter of considerable academic debate. Heidegger's work has, however, become an integral part of many, very different political philosophies and of a very necessary and fundamental self-critique of the Western philosophical tradition. In this sense, it needs to be employed cautiously and critically but cannot be ignored.

a narrative that warns of our own existential insignificance, our impending approach toward "the last syllable of recorded time" (Shakespeare 1953, 553). *This* recuperation, however, no longer suggests the universal inclusivity of human semiosis, the God-like ability of human language (Benjamin 1991; Smith 2001) to express everything that matters. Rather, it signifies that the earth far exceeds our abilities to hold it, momentarily frozen, in the halting flows of words and images. Moreover, despite modernity's "progressive" disenchantment of the world (Weber 1964), the earth can still, on occasion, encourage us to take everything that then appears and affects us as a significant moment in the changing contexts of our lives. *In this materio-semiotic sense, the earth might, perhaps, be considered insistently providential, its intrusions "telling,"* although what we are told about our own situation is often far from comforting.

Profound experiences of earthly indifference, epitomized by climate change and instantiated in fracturing and melting glaciers, raging wildfires, pests, and pandemics, are the new abnormal. In this sense, what the iceberg might tell us is that assumptions of theological providence and its successor, progress, are both untenable. If this prodigious event is not a message for humanity from God, it is certainly not a mark of human progress either. These anthropocentric imaginaries can no longer, if they ever did, provide the overarching ideational and material context of our lives, nor can they any longer underpin our presumptions about our place and future here "on" (or rather as composing a temporary part of) earth. Yet, despite this unpredictability, the earth still *provides,* and for far more than just humans, in far more, albeit sometimes uncanny, anarchic, weird, unfathomable, and episodic ways. If the world is no longer a matter of divine providence, of design and intent to succor humanity, nature as such still offers other forms of provisional ecologies that some can continue to inhabit. Alienated or not, we are all, and all remain, earthlings.

Progress, Providence, and the Anthropocene

DISCOURSE AROUND THE ANTHROPOCENE often forces nature to appear as little more than a surrogate for capitalism, both "systems" being destructively and irreversibly entangled, both now deemed impervious to the sufferings they inflict. Indeed, the Anthropocene seemingly denotes a hopeless situation for all, except those retaining an unshakable faith in technical fixes or a monomaniacal desire to profit from others' misery (Klein 2007). There is certainly no shortage of such people, yet to describe a situation where the entire world is threatened by an economic system inducing mass extinction and climate change as either "providential" or a matter of "progress" now seems entirely implausible.

Providence is, in any case, a term that has largely vanished from all but theological lexicons. Yet progress *is* still a world-structuring concept—or perhaps, more accurately, it remains, *for the moment,* a key aspect of a world-shaping *imaginary,* by which we mean the shared and circulating materio-semiotic flows that (de)limit possibilities for visualizing, expressing, and understanding our social-ecological situation.[1] Science, for example, is still, by and large,

1. Imaginary in this sense is not simply a reference to the arrangement and emergence of patterns of human thought understood in terms of an individualized psychology, still less to imagination as it is often op-

envisaged as a process of continual epistemic progress despite its contributions to, and expression in, the rather obvious downsides of "inventions" like nuclear weapons, virulent pesticides, nanoparticles, plastic waste, and mass surveillance. The point here is not to argue whether science per se is either "good" or "bad" but to illustrate how impossible it has been to engage in any such debate without immediately invoking an image of its contribution to something called "progress," even if only as an overarching alibi to justify that which is ethically inexcusable.

We might say, then, that an imaginary, such as that surrounding progress, constitutes a pervasive *atmosphere* of possibilities within which a given society operates, one inhaled and exhaled through every interpretative, creative, productive, and destructive act, even when dreaming of alternatives. In this sense, this imaginary also denotes a hermeneutic "breathing space" that, so long as it is materially and ideologically sustainable and sustaining, continues to *inspire* that society. Ironically, with the advent of global climate change, it is increasingly clear that *the imaginary associated with progress no longer has a future* (Beradi 2011, 17). The atmosphere it has produced is, quite literally, becoming unbreathable, its waters anoxic. Have we finally, then, reached the end of progress? If so, what might "end" here signify: the end of an illusion (Gray 2004), the end of a colonizing European ideology (Allen 2016), the end of a/the world (Danowski and Viveiros de Castro 2017), and/or, perhaps, the completion of the, rather unexpected, telos of modernity? What

---

posed to "reality." Rather, it has close connections with Anderson's (1991) notion of the nation-state as an "imagined community" that is materially influenced and enacted. It might also be linked with Castoriadis's (1998, 146) notion of the "actual" and "radical" imaginary "manifested indissolubly in both historical *doing* and in the constitution before any explicit rationality, of a universe of *significations,*" and a materio-semiotic extension of the "social imaginary" defined by Taylor (2005, 23) as "that common understanding that makes possible common practices and a widely shared sense of legitimacy."

will the sublimation of this imaginary leave behind when everything that once declared itself so solid is revealed as so much hot air?

In considering the immanent end of this world-shaping and shaped imaginary, we might do well to return to its origins to better understand how "progress" became insinuated into every aspect of a synonymously "*modern*" epoch. Lloyd (2008) carefully traces how, from the seventeenth century onward, "progress" actually emerges as a replacement for theological (and, before that, much more ancient) notions of "providence." It expresses an increasingly secular but, if anything, even more narrowly anthropocentric idea of agential teleology. She tracks this movement philosophically (although we would also emphasize its immense materio-semiotic resonances and ramifications) from ancient Greek tragedy through Augustine, Descartes, Spinoza, and Rousseau to Kant, where this change and emergent imaginary first become explicit.

Kant's argument is subtle but far-reaching. He admits that we *could* just interpret natural relations in terms of how their existence assures the *provision* of sustenance for humans. For example, we might claim that animals are placed here by God for us to domesticate, eat, and so on. Or we could interpret the world *purposively,* for example, "herbivores are there to moderate the opulent growth in the plant kingdom, which would otherwise choke many species of plants." This approach leads to a similarly anthropocentric, but not identical, conclusion that "man [*sic*] is there to hunt the predators [of these herbivores] . . . and so establish a certain equilibrium between the productive and destructive forces of nature" (Kant 1987, 427). The former possibility is clearly providential; the latter might seem to link nature's own purposiveness with a hierarchic role for human stewardship over the earth, though, from Kant's perspective, its problem is that it opens the door to reducing the "dignity of [humanity's] being a purpose" to having only what we might now consider a naturalized ecological and functional role, where humanity "would only hold the rank of a means" (Kant 1987, 427) rather than a "final end." That is, humans would become just one species among many, each having its naturally appointed role in the "great system of the purposes of nature" (Kant, as cited in Zuckert 2007, 128).

This phrase illustrates how Kant thinks it is acceptable, up to a point, to employ a principle of purposiveness in understanding the natural world, but he deems this employment "*merely* reflective not determinate" (Zuckert 2007, 130). In other words, we may sometimes find this attribution of purposiveness necessary as a way of thinking about "ecological" relations, but this does not actually *explain* those relations. Kant argues that we actually have an "obligation to give a *mechanical* explanation for all products and events in nature, *even the most purposive,* as far as it is in our capacity to do so" (Kant, as cited in Zuckert 2007, 130, original emphasis). Thus, for Kant and those who follow, enlightened scientific approaches should seek to excise both purposive and providential explanations from their accounts of *nature.* The language of purpose should really be reserved for human activities.

The centrality of human exceptionalism to his philosophical project certainly provides one reason why Kant is unwilling to reduce humanity to its "ecological" role in nature's great system. However, he also argues that a more thoughtful and thorough acquaintance with the earth's "natural history" shows that what looks at first sight like the purposive ordering of natural beings and the habitats in which they seem to thrive is actually wholly unintentional and, what is more, is produced by "causes that are more likely to be devastating than to foster production, order, and purposes" (Kant 1987, 428).

> Land and sea contain memorials of mighty devastations that long ago befell them and all creatures living in or on them. Indeed, their entire structure, the strata of the land and the boundaries of the sea, look quite like the product of savage, all-powerful forces of a nature working in a state of chaos. (Kant 1987, 428)

In other words, seen in the whole, there is no overarching purpose, order, or preordained "balance of nature" there to be maintained, and Kant (1952, 108) concludes that "without man the whole of creation would be a mere wilderness, a thing in vain, and have no final end." Kant, then, wants to argue that "Man is the ultimate purpose of creation here on earth" (Kant 1987, 427) but needs to do this without reverting to the naive, selective, and dubiously positive notion of theological or natural providence he has just critiqued. His solution is to emphasize human "progress." The antinomies of nature, that is, its apparent provision of sustenance *and* its often chaotic and destructive activities, are to be taken together and subsumed by and within the struggle of humanity to progressively understand, and gain mastery over, the natural world. In this imaginary, what harms us individually may actually help humanity, over time and as a whole. "Man wishes concord, but nature, knowing better what is good for his species, wishes discord," for without this discord, humans, "as good natured as the sheep they tended, would scarcely render their existence more valuable than that of their animals" (Kant, as cited in Lloyd 2008, 288).

In Kant's philosophy, nature's purposiveness is ultimately reduced to *providing* the sometimes challenging circumstances necessary to force humans to develop their latent potential to assimilate nature to their own purposes and ends. We might even say that this ultimate purpose is ethico-politically characterized in his *Idea for a Universal History with a Cosmopolitan Purpose* (Kant 1991).[2] Indeed, the third proposition of this work describes this process in terms of nature's "intent" and its human supersession in what Apel (1997, 97) refers to as a "quasi-dialectical idea of a 'cunning of nature'":

> Nature has willed that man should produce entirely by his own initiative everything that goes beyond the mechanical ordering of his animal existence, and that he should not partake of any other happiness or perfection than that which he has procured for himself without instinct and by his own reason. . . . It seems as if nature had intended that man, once he had finally worked his way up from the

2. Which is not to say that this work constitutes a conscious design for Kant's philosophical project; rather, along with his later work *Toward Perpetual Peace,* this is as near as we get to a this-worldly ethico-political characterization of humanity's purposive and progressive possibilities.

> uttermost barbarism to the highest degree of skill, to inner perfection in his manner of thought and thence (as far as is possible on earth) to happiness, should be able to take for himself, the entire credit for doing so and have only himself to thank for it. (Kant, as cited in Apel 1997, 90)

How generous of nature, to take no credit! To adopt a modern vernacular, it seems that providence, in the form of nature, does not guide history to its final ends but rather provides "opportunities" to be exploited by human entrepreneurs, self-made men and/or Man, who then take(s) *all* the credit and reap(s) *all* the profit. Here we can see how Kant anticipates the movement from a providence that is weakened but not entirely expunged to a progress that effectively becomes a rationalist and humanist equivalent of a theodicy of good and evil, one that becomes transformed, elaborated, and materio-semiotically enacted in innumerable (and often incredibly destructive) ways as it inspires the imaginary of an emerging modernity.

This raises two important issues. First, Kant's envisaging this specifically European solution to a specifically European philosophical problem as the "universal" end for humanity and the earth does not necessarily lead to cultural tolerance, as the use of the term *cosmopolitan* might now suggest, or to "perpetual peace." Rather, in aligning progress with European developments, it actually provides another justification for European colonialism and its "overcoming" of the "challenges" posed by very different (purportedly primitive) peoples and cosmologies that are deemed to stand in the way of that progress to "inner perfection." Second, because nature's ultimate purpose is now entirely focused on humanity as both an end in itself and for the earth as a whole, nature's value is confirmed as being entirely instrumental—it is just a means to that end; it cannot, for Kant, be a matter of *ethical* concern. Nature as such, we might say, also becomes a victim of modern Eurocentric indifference. As we shall see, this leaves the very idea of a final end, and the possibility of its being distinguished from nature's great system of purposes, resting on the unstable ground of human exceptionalism.

Kant's optimism regarding how anthropocentric agency overcomes natural chaos and intransigence exemplifies the "progressive" and anthropogenic imaginary that comes to infuse everything from Western culture to capitalism, Marxism to meteorology. Golinski (2007), for example, describes how the temperate but changeable nature of the British climate came to be regarded as a natural condition that could actually explain the origins of British innovation and progress. Providential accounts of weather had completely dominated the seventeenth and early eighteenth centuries, but an emerging science of meteorology sought to provide "enlightened" accounts both of this "providential regularity" (Golinski 2007, 65) *and* of extreme departures from these norms. Events like the Great Storm of 1703 and the summer haze of 1783 were obviously, as in Kant's examples of past devastations, *providentially* challenging. The new meteorological explanations accounted for this through an increasingly secular science that both removed requirements to second-guess God's purposes and, through their repetitive gathering of weather data, sought to replace a temporality of discontinuous weather events—a sacred *kairos*—with a continuous secular climatic *chronos* (Golinski 2007, 78; Serres 1995). This emphasis on long-term weather patterns (climate) offered the prospect of reliable predictive capacities and even an eventual managerial mastery of climate. British colonists, for example, claimed that their environmental "improvements," clearing forests and draining swamps, were also changing the climate of colonized lands for the "better." "They expressed the optimism of the Enlightenment in their conviction that the American climate had been tamed by human enterprise and reason" (Golinski 2007, 5). *Anthropogenic climate change has clearly been part of the modern Western imaginary for longer than we might imagine!*

"Progress," then, might be understood rather differently as that movement in every field away from an imaginary of a providential nature toward a nature that humans must struggle to manage and subsume within their own, consciously directed social projects. *Humanism, in this sense, becomes the explicit ideology of the imag-*

inary of progress. This change, from understanding nature as the material expression of God's *providential* concerns for humanity to nature as an infinitely transformable resource for humanity's *projects* (as standing reserve, in Heidegger's [1977b] terminology) has many implications, yet it obviously retains a core vision of human exceptionalism while dramatically reducing the imaginary possibilities for involvement with and appreciating nature. Indeed, as we enter the so-called Anthropocene, an epoch dominated by anthropogenic processes, many claim we have erased nature altogether (Morton 2007; Žižek 2008).[3] Even to refer to nature now risks being labeled theoretically naive, obsessively nostalgic, and perhaps, because it is not deemed sufficiently "progressive," politically reactionary. For nature, as Lefebvre (1994, 31) somewhat disdainfully remarked, has been left behind: "anyone so inclined may look over their shoulder and see it sinking below the horizon behind us . . . lost to thought . . . defeated" and waiting only "for its ultimate voidance and destruction." The progressive humanization of nature is, it seems, nearing completion.

How shocking, then, that climate change, with melting glaciers, raging wildfires, droughts, floods, and mass extinctions, confounds such speculations on a global scale. Of course, some simply deny the reality of climate change, while many more remain in a state of denial regarding its implications. Whether one regards such denial as existential "bad faith" or just a spur to "theoretically enterprising" defenses of humanism depends on whether one is willing to continue buying cut-price retreads of worn-out revolutions: scientific, agricultural, American, industrial, proletarian, informational—

3. For a more detailed critique of Morton on "nature," see Smith (2011). Žižek (2008, 445), like Morton, suggests that "what we need is an ecology without nature: the ultimate obstacle to protecting nature is the very notion of nature we rely on," but this is disingenuous, because the fact that other species and entire ecological communities are imperiled by climate change does not seem to concern him at all. Žižek is interested specifically in disputing any idea of a "balance of nature," describing nature, very like Kant, as a "mega-catastrophe."

modernity is littered with them. All once so "promising," all now climatically/climactically compromised. Only this state of denial allows climate change to appear in a Kantian light, as just one more challenge for humanity to overcome.

No doubt, even at the height of meteorological optimism, it was always the case that "strange weather phenomena showed the natural world in its most recalcitrant aspect, continuing to resist attempts to bring it within the pale of scientific reason" (Golinski 2007, 76). Now, however, it is scientific reason that propounds the world-encompassing nature of a new climatic normlessness. Even meteorology no longer predicts climatic regularities but rather foresees only intemperate and increasingly extreme weather events. *It is not just ironic but tragic that the imaginary of progress, so deeply and materially implicated in subverting the climatic regularities previously deemed providential, now leaves us facing an anthropogenic realization of Kant's intimation of "nature as chaos."*

And so, climate change now appears as something apocalyptic, potentially creating a "final end" of humanity and the world in a very different sense from Kant's intentions. Indeed, we might say that the emergence of climate change is to "progress" what the Lisbon Earthquake of 1755 was to "theological providence" (Molesky 2016). Whereas the emerging humanist imaginary of modernity would eventually come to reenvisage the Lisbon Earthquake as a *natural* disaster, rather than an act of God's mysterious purpose, no comparable shift or excuse is yet available for climate change. Indeed, insofar as the Anthropocene denotes the "success" of modernity's assimilative project, the humanization and the consequent "end" of "nature," it also exposes the abject failure of the imaginary of progress to produce any kind of a future that remotely resembles a livable world where all the earth's inhabitants are provided for.

Gaia as an Incipient Terrestrial Imaginary

THE LOSS OF a providential imaginary is *not,* then, as it is so often portrayed, an inevitable outcome of historical and epistemic "progress" (reified as a transcendent teleological process that replaces superstitious ignorance with knowledge) so much as it is a by-product of a geographically and historically specific imaginary sold on (the basis of) this historicist story. It is not, of course, *just* a story—as an imaginary, it has a key constitutive role in the materio-semiotic production of modern spaces; it is expressed in concrete, plastics asphalt, nanoparticles, microelectronics, and so on.

Yet even as climate change materially deconstructs "history as progress," we might consider how the imaginary of a provident earth has not been entirely lost, nor has nature as such actually been confined, as Lefebvre suggests, to history's dustbin. Both survive despite capitalism, technology, and humanism having "conspired" to continuously erode, reduce, and restrict the materio-semiotic breathing space required for these modes of expression to inspire alternative imaginaries—declaring them regressive, off-limits, vacuous, and void. Indeed, if progress was itself a deferred providentialism, where human ingenuity and labor were supposed eventually to provide everything we could possibly need, then now, here in the so-called Anthropocene, it is this progressive imaginary itself, once so inspirational, that is rapidly running out of oxygen. Its last gasps continue to invoke a world purposefully organized by and for humanity, and yet any sense of humans constituting a "final end" in anything like Kant's sense appears increasingly illusory. The emergence of speculative forms of hyperhumanism, whether of the right (e.g., the Breakthough Institute) or "left" (e.g., accelerationism), are just further extrapolations of an anthropocentric ecological indifference in denial of the failure at modernity's heart (see Danowski and Viveiros de Castro 2017). To adapt Joe Hill's words, they offer more pie in the sky as things die. To recall Herzen (Ward and Goodway 2003, 86), "an end that is infinitely remote is not an end, but a trap."

These issues, of providence lost (Lloyd 2008) and the *lack* (the existential hollowness) of progress, are actually central to recent critiques of anthropocentrism associated with the advent of climate change. For example, the (often vitriolic) arguments around Gaia, Lovelock's (1979, ix) "shorthand for the hypothesis . . . that the biosphere is a self-regulating entity [a system] with the capacity to keep our planet healthy," are vitally important precisely because they implicitly (and sometimes explicitly) appear to offer *the prospect of new ecological imaginaries,* of potential breaks with, or evolution from, regimes of human progress and theological providence (Latour 2017a, 2017b).

Gaia theory offers a vitally important opportunity to reflect upon the prospects of an imaginary where different, relatively nonanthropocentric understandings of earthly indifference proliferate. James Lovelock, who, along with Lynn Margulis, developed the Gaia hypothesis, has long argued that the inflated estimation of humanity's own central importance is about to suffer a potentially fatal blow with climate change, one that will put us in our "proper place" not as "owners, managers, commissars, or people in charge" but just as "part [and, it seems, an entirely expendable part] of the Earth system" (Lovelock 2009, 6). In Gaian imagery, the earth is a homeostatic whole, a "vast being who in her entirety has the power to maintain our planet as a fit and comfortable habitat for life" (Lovelock 1979, 1). Humans are explicitly *not* the final end of this planetary entity; we are just a recent addition in its evolutionary history and one now seriously compromising its capacity to maintain life in its current forms. These activities may not threaten Gaia herself, to whom Margulis (in)famously refers as "a tough bitch" (as cited in Lovelock 2006, xii), but certainly have dire consequences for many species and most, if not all, humans. In Margulis's words,

> *the pandemic we call progress* (e.g. deforestation, desertification) are for Gaia, only petty activities. . . . Gaia continues to smile: Homo sapiens, she shrugs, soon will either change its wayward ways, or, like other plague species, will terminate with a whimper in the current scourge, in this same accelerated Holocene extinction it initiated. (Margulis, as cited in Harding 2006, 12, emphasis added)

These Gaian critiques of progress are often conjoined with arguments against the anthropocentrism underlying the instrumentalist reduction of nature to "standing reserve." As Lovelock (2006, 135) notes,

> the concept of Gaia, a living planet, is for me the essential basis of a coherent and practical environmentalism; it counters the persistent belief that the Earth is a property, an estate, there to be exploited for the benefit of humankind.[1]

Contra Kant, then, Gaian earth is not a chaotic wilderness but an autopoietic (self-creating) entity that currently includes, but is not governed by or for, humanity. Kant's concerns about humanity losing the "dignity of being a purpose" are made manifest in Gaia, which is agential, is nonanthropocentric, and undermines overarching narratives of progress.[2] In this sense, it might indeed seem

1. Lovelock (2006) here makes a distinction between the Gaia hypothesis as originally formulated and Gaia theory; the latter emphasizes that it is the whole earth, and not just the totality of its *living* components, that constitutes a homeostatic system.

2. Although, in his typically inconsistent manner, Lovelock sometimes suggests that humans may indeed have exceptional, unique, and potentially overarching planetary *purposes*, including being "stewards." For example, he muses whether we might function as a "Gaian nervous system and brain

an inspiration for new imaginaries that could veer away from the impassive acceptance of the "Anthropocene" as a planetary telos, and this is surely Latour's (2017b) hope, as his recent paean to Gaia demonstrates. Here Latour speaks of the need to explore an ecological crisis that is engendering a "profound mutation of our relation to the world" (8) and also to rethink "the idea of progress" (13)—not so much as the linear forward movement modernization promotes as its self-image but as running away from and destroying the "so-called 'shackles of the past'" (Stengers 2018, 100; Latour 2018).

Gaia theory also appears to offer a break with notions of final purpose and providence. Lovelock (2000, 11) claims that Gaia

> is an *evolving system* . . . [where] the self-regulation of climate and chemical composition are *entirely automatic*. Self-regulation *emerges as the system evolves*. No foresight, planning or *teleology* . . . are involved. (emphasis added)

Latour (2017a, 133), quoting this passage, says it "would be hard to be clearer about the absence of Providence." Indeed, Latour (2017b, 69) is very explicit in declaring that Lovelock is

> offering us what I take to be *the first totally non-providential* and non-holistic version of what it is to compose a whole. Gaia, in spite of her godly name, inherits none of the political theology that has paralyzed nature as well as evolution. (emphasis added)

Whether this break with progress, providence, and purpose is as final as Lovelock and Latour sometimes suggest is doubtful.[3] Lovelock (2000), for example, though not uncritical of scientific priorities and science's resistance to new ideas, retains a positivistic faith

---

which can consciously anticipate environmental changes" (Lovelock 1979, 147) and might potentially use our technologies to save the planet from things like meteorite strikes.

3. Note, for example, the reoccurrence of previous tropes, when Latour (2017a, 291) imagines us exploring, through Gaia, "in a way, a path for *progress* and *invention*" (emphasis added).

in value-free science as epistemic progress. (This leads him, for example, to castigate Rachel Carson for having been too political [Lovelock 1979, ix], while later advocating the ceding of political power to a scientific elite to cope with a state of climate emergency [Lovelock 2006, 153].) Latour (2018, 81), while rejecting the ideology of progressive modernization and its continually receding horizon, explicitly reintroduces a notion of progress toward his own version of a more *Down to Earth* third way:

> Celebrating the forward march of progress cannot have the same meaning when one is heading towards the Global as it does when one is heading toward "decisive advances" in taking the Earth's reactions to our actions into account.

Similarly, despite his blunt rejection of any teleological associations, Lovelock's (1979, 11) early iterations of the Gaia hypothesis certainly did employ a language of purpose, making claims like that Gaia "*seeks* an optimal physical and chemical environment for life." This language of purpose (and even providence, where maintaining an environment for life, and not just humanity, is concerned) was a target for many of his early scientific critics (see Ruse 2013). In later writings, as climate change becomes less speculative and Gaian ideas are taken up in Earth systems science, the language of purpose and of providence is more qualified. Its use is explicitly justified as a metaphorical aid to visualizing and reconceptualizing the world, a communicative gloss that overlies the science and not something implying an explanation dependent on a planetary telos.

This appeal to communicative efficacy is also how Lovelock (2000, 12) justifies his references to the earth as a "living organism" and the earth's anthropomorphic portrayal in the figure Gaia herself, that conjunction between mythic mother and cybernetic planet—*lending* itself to poetic metaphor and *being* "hard science." Gaia's association with an ancient earth goddess has, Lovelock recognizes, been central to its popular appeal *and* to surrounding controversies. And, although he has no intention of relinquishing her, when pressed, Lovelock often suggests that these metaphors are *scientifically* indefensible and, so far as he is concerned, ultimately unnecessary:

> When I talk of Gaia as a superorganism I do not for a moment have in mind a goddess or some sentient being. I am expressing my intuition that the Earth behaves as a self-regulating system. (Lovelock 2000, 57)

Gaia's anthropic figure is, paradoxically, also at the heart of his attempts to rhetorically undercut the idea of humanity as the ultimate telos for life on earth. Gaia is not only Margulis's "tough bitch" smiling down at anthropogenic devastation but, as Lovelock (2006, 147) similarly declares,

> we now see that the great Earth system, Gaia, behaves like the other mythic goddesses, Kali and Nemesis; she acts as a mother who is nurturing but ruthlessly cruel towards transgressors, even when they are her own progeny.

The gendered portrayal of the earth as nurturing mother/goddess turned vengeful harpy is obviously troubling for any feminist analysis (see Roach 2003; Sands 2015), but *The Revenge of Gaia* is a trope not confined to Lovelock (2006) and Margulis (see, e.g., Liotta and Shearer 2007) and is widely echoed in recent responses to events like Covid-19, although ridiculed by Latour (2020). Lovelock may just be exploiting anthropomorphic tendencies to communicate his ideas, but to portray Gaia switching between care and ruthless cruelty, to refer to "her" impending revenge, now served climatically hot, hardly fits with the cybernetic science that Lovelock claims underlies his image, a science that, as he frequently points out, actually regards the earth as systemically *indifferent* to humanity's fate. In this sense, the earth can never have been either nurturing or vengeful; rather, it simply lacks any capacity to allow us to describe it in any such way. This certainly reflects the assumptions underlying scientific objectivism. Should we then just dismiss these anthropomorphic attributions as purely rhetorical? In what sense, if any, is Gaia actually caring, heartless, or indifferent?

As we have seen, for Lovelock, the answer seems to be that there is indeed a gap between scientifically described reality and communicative rhetoric, Gaian science and Gaian metaphor:

> You will recognize, I am continuing to use the *metaphor* of "the living Earth" for Gaia; but do not assume that I am *thinking* of the Earth as alive in a sentient way, or even like an animal or a bacterium. . . . It has never been more than an *aide pensée*. (Lovelock 2006, 16, emphasis added)

However, only a few lines later, we are told that to understand that,

> global climate change requires us to *know* the *true nature* of the Earth and *imagine it* as the largest living thing in the solar system, not something inanimate like that disreputable contraption "spaceship Earth." Until this *change of heart and mind* happens we will not *instinctively sense* that we live on a planet that can *respond* to the changes we make, either by cancelling the changes or *cancelling us*. (17, emphasis added)

This passage is both fascinating and dramatically ironic. In direct contradiction to his previous statement, it elides gaps between ontology, knowledge, emotion, and imagination, between truth and metaphor. Lovelock (2000, 17) is surely right to assert that "metaphor is important," that it plays an important role in science (18), *and* that there are important semio-material differences between imagining the earth as a mechanical spaceship, a "contraption," and as a living organism. But why this difference *unless* this metaphor draws on a residual understanding of nature as opposed to human artifice?

> [Gaia] is an alternative to that pessimistic view which sees *nature* as a primitive force to be subdued and conquered. It is also an alternative to that equally depressing picture of our planet as a demented spaceship, forever travelling, *driverless and purposeless*, around an inner circle of the sun. (Lovelock 1979, 12, emphasis added)

The strength of Lovelock's rejection of this "inanimate" and "disreputable" metaphor "spaceship Earth" certainly suggests that

Lovelock finds this mechanistic analogy demeaning. In Latour's (2017b, 77) words,

> Gaia, for Lovelock, could be called No Machine and that's why, of all the metaphors he criticizes, none is damned more relentlessly than Spaceship Earth.

However, if this is so, then it is doubly ironic that Lovelock's own preferred solutions to climate change all involve upping the ante in terms of massive technological interventions like nuclear power and geospatial engineering, that is, actively diminishing any gap between "spaceship Earth" and "living Earth." His earlier separation between Gaian image and science also ensures that he often remains captured by the same objectivizing frameworks that deny Gaia existence in anything but its most reductive, mechanistic form, as Earth systems science. This inevitably finds expression in Lovelock's own language. For example, his claim above that the earth might respond to our activities by "canceling us" seems strangely detached and calculating, more like the actions of that spaceship computer HAL in Kubrick's *2001: A Space Odyssey* than the earth we actually inhabit, and still less the actions of a "goddess."

More importantly, the systems theory underlying Lovelock's *thinking* about Gaia (initiated through his work with NASA!) is itself the very form of technologically derived paradigm that has dominated attempts to connect (and "manage") social, mechanical, informational, and natural "systems" over recent decades. In its dominant forms, cybernetics epitomizes the image and imaginary of progress in proclaiming that these "systems" work according to the same fundamental principles, all expressing the same underlying systemic "truth" of the world—earth as planetary system. From its inception cybernetics has, quite self-consciously, regarded itself as a way of eradicating ontological difference between nature and artifice. Weiner's massively influential book originally defining the field was, after all, entitled *Cybernetics; or, Control and Communication in the Animal and the Machine* (1948). Adopting this perspective obviously makes it all the more difficult to distinguish between spaceship Earth and living Earth,[4] a distinction now further blurred in Lovelock's (2019) prediction of an imminent age of "hyperintelligent" cyborgs he names the Novacene.

Now Latour rightly warns against thinking of cybernetics as a scientific truth underlying Gaian images. To "speak of the earth as a 'system' is just as confusing, because . . . its political and philosophical pedigree is much harder to render explicit" (Latour 2017b, 62); that is to say, systems theory is *often objectifying* but is in no sense a *neutral* way of describing the earth. Yet this is just what Lovelock does when he attempts to parse the politically and emotionally engaging *image* of Gaia and what he regards as the objective scientific reality of the earth. Latour (2017b) himself ridicules one of Lovelock's scientific critics, Tyrrell, for ascribing to Lovelock the idea of "Earth as thermostat," but that "metaphor" *is,* after all, Lovelock's, and Lovelock does think that thermostats, humans, and the earth *are* all examples of self-regulating systems—he says so! We surely have to ask whether this is really any less "disreputable" (Lovelock 2006, 17) or "depressing" (Lovelock 1979, 12) than speaking of the earth as a spaceship?

Such tensions and inconsistencies are palpable in Lovelock's work. They are, in part, what makes it so controversial and also arise, as Lovelock and Latour recognize, from the need for a more-than-just-scientific account of the earth to gain political traction and *overcome our systemic indifference to climate change,* to effectively engage both "heart and mind." To this end, Lovelock seems willing to employ the figure of Gaia as a kind of ventriloquized voice for

4. Contrary to the popular phrase, there is a sense in which cybernetics, at least in its earlier instantiations, was, quite literally, rocket science, although it is also now information science, meteorology, economics, ecology, and so on. There are, of course, developments in systems theory that provide opportunities to reconceptualize our earthly relations, as Bateson's (1987) work and those strands of second-order systems theory emerging from the work of Maturana and Varela exemplify (Wolfe 1998; Clarke 2020; Young 2020). These, however, are not the forms of systems theory Lovelock employs.

the earth, a rhetorical tool with popular appeal, useful in drawing attention to the unfamiliar underlying science.

Latour, however, takes a rather different approach. He is clearly captivated by the very way that Gaia as a figure entangles and conjoins science, politics, and a certain vague religiosity. Gaia, for Latour, is a theoretical and earthly *incarnation* of "nature/culture" that exemplifies his famous claim that "we have never been [and can never be] modern" (Latour 1993). That is to say, we have never actually succeeded in severing politics or religion from science, culture from nature, metaphor from fact. In this sense, while both agree that Gaia is not actually a goddess, Latour regards her as far from being *just* an *aide pensée,* because this kind of admixture characterizes his advocacy of a current requirement to ecologize rather than modernize (Latour 1998). Latour (2017b, 76) agrees that Lovelock "remains a totally naive believer in mechanical philosophy and his politics is equally naive and often counterproductive" but often excuses what Lovelock (Latour 2017a, 94–98) has *actually* said, because he considers that Lovelock is genuinely experimenting with trying to communicate Gaia theory against ingrained modernist biases.[5] Indeed, Latour (2017b) even goes so far as to suggest that Lovelock's switching between radically different metaphors is a deliberate tactic to resist reductive interpretations. As our discussions suggest, this seems extremely unlikely. However, it seems fair to say that Latour finds Gaia so conducive precisely because it begins to enunciate an incipient terrestrial imaginary, a different way of coming to address the earth.

5. Latour castigates Tyrell for inattention to the subtleties of Lovelock's language but entirely overlooks these "subtleties" when it suits him. For example, immediately following Lovelock's denial of teleology quoted earlier, Lovelock (2009, 6) says, "The Earth has not evolved *solely* for our benefit" (emphasis added). But how does a reader "sensitive to the tropisms of prose" (Latour 2017a, 132) interpret this? Might it not be suggestive of a form of more-than-just-human providence?

The Imaginary End(s) of the World

SO GAIA MAY BE MUCH MORE than just an *aide pensée,* but to say this does not yet begin to elucidate an earthly imaginary or resolve the apparent contradictions in Lovelock's account of Gaia. Lovelock's insistence on the contrast between "living Earth" and "spaceship Earth" does, however, suggest that we need to think more deeply about how the earth might, or might not, come to be understood as purposeful and/or providential.

Insofar as he is mechanistic, Lovelock seems to evacuate the earth of all purpose (and therefore any association with an imaginary of theological providence or overarching progress), reducing living nature and human life to nothing more than the cybernetic systems that together compose the whole earth. That said, his distinction between spaceship Earth (qua inanimate contraption) and living Earth makes it absolutely clear that he wants to resist at least some of the implications of mechanistic approaches. Describing the earth as an actual machine is deemed inaccurate, explanatorily limiting, and degrading.

Ever since his NASA days, Lovelock has considered that life is distinguished from nonlife by its homeostatic capacities; living beings can, he argues, be defined as "self-regulating systems." The earth, as we have seen, is also a self-regulating system involving living beings, but one that exceeds any comparison to a single organism, be it animal, bacterium, and so on. He also, as we have seen, recognizes that some "machines" are *not* living organisms, nor are they the planet Earth, but, like a thermostat, are nonetheless self-regulating systems. It seems to be the case, then, that from Lovelock's perspective, being a self-regulating system is necessary, but not actually sufficient, to define a living being or the earth. It also seems impossible for all this to make sense *unless* Lovelock holds that there is a distinction between nature and artifice and that living beings and the earth as natural entities are actually deemed to recuperate, in some unspecified way, a self-generated nonteleological "purposefulness," rather than just being incorporated in, and made subject to, the "external" purposes of human designs (i.e., "foresight" or "planning"; Lovelock 2000, 11).

In other words, despite recognizing that as a "good (post-Kantian) scientist," he should have avoided ever describing the world in terms of purpose, Lovelock's mode of addressing the earth often reveals quite different feelings. Remember that the problem from a Kantian perspective is that a nonteleological purpose is not really a purpose at all. All "genuine" purposes are deployed to reach toward final end(s), and these ends both are necessarily anthropogenic and progress toward the full realization of humanity's potential. Other attributions of purpose are just "reflective" misapplications of the term, an error that science will eventually explain away, leaving only human (and theological) purposefulness remaining. As we have seen, Kantian purposefulness clearly expresses the kind of exceptionalism that regards humans as being in a unique position to provide an "external" purpose to objects; indeed, it limits the very notion of purpose per se to the way an object or artifact serves as the means for human projects to succeed. The agency here is all human, the final end is projected by humans, and nature is just a more or less refractory raw material. This is how the progressive imaginary understands the human condition. It is expressed in the celebration of human ingenuity, the Marxist reification of labor as the active agent of history, the commodification and mobilization of every aspect of the world to achieve economic growth, and so on. This is a description of the progressive imaginary *in its own terms,* moving continually on from project to project, regarding everything that no longer serves its previously defined purpose or has passed its sell-by date as a useless (unless recyclable) waste product.

Now Lovelock's "driverless" spaceship is "purposeless," but his earth is deemed purposeful independently from, and prior to, any human purposes. Contra Kant, and given Lovelock's evolutionary perspective, human purposefulness has to be *derivative* (from nature), and it is *not determinate*; there are no *final* ends. In this sense, to conflate the earth with a spacecraft is indeed "despicable" (the moral tone is always present in Lovelock's rejection of this term) because it only exhibits our obsession with our own technical accomplishments, the paucity of our imaginations, and the dire, and self-destructive, limits of an anthropocentric imaginary that is actually turning a "living" autopoietic world into waste material. By comparison with the living Earth, a driverless spacecraft just orbiting the sun in the earth's place would soon lose any purpose it ever had bestowed upon it. It would, after a few revolutions, be no more than one more piece of space junk, like Elon Musk's automobile propelled on an endless road to nowhere, a space-suited dummy propped behind the steering wheel—the perfect image, and unfortunately the *reality,* of capital-driven egoism and the failing imaginary of progress. For Lovelock, the earth is *not* an object, and its purposes are internally generated.

We need, though, to follow this idea of purposefulness further. This does not require entering detailed arguments about Aristotelian understandings of *telos, arche, technē,* and *physis,* although Heidegger's (1993, 1998) own discussions of these mutually imbricated concepts does offer a critical opening here. Heidegger's later work carefully unpacks the way efficient causes come to set the (mechanistic) standard for defining causes in general to such an extent that "we no longer count the *causa finalis,* telic finality, as causality" (Heidegger 1993, 290). That is to say, historically (as Kant illustrates), we have come to regard final ends as playing no part in nonhuman nature. Yet Heidegger allows us to go further, suggesting that the very idea of a final end, that is, a purpose that gives overarching meaning to things explained as having preceded and been directed toward it, is itself an integral aspect of how the ancient Greeks and those under their influence began to think "technology." *We might even say that the very idea of a final end is itself a reflective product of taking a specific and limited "technological" mode of relating to the world as the "model" for Being itself.*

This point is crucial but requires further explication. First, note that for Heidegger, technology is not at all a specific array of technical machinery, or even technics in general, but a mode of engagement that transforms the world through challenging and ordering it. It is a specific mode of addressing the earth where the earth is forced to appear as material for human projects. Also, for Heidegger, technology is not something superimposed on the world by a (human) being who stands apart from the world; it is an instrumentally oriented occasioning of our being-in-the-world. *Technē,* from its ancient Greek origins, refers to crafting, making artifacts for a particular purpose, an end, like a pot or, in Heidegger's (1993) example, a chalice—but a spaceship would do just as well. This end, this *causa finalis,* seems to be all too readily identifiable in such cases, albeit also intimately connected to the other aspects of the artifact's production, namely, its intended form, the material employed, and its efficient causes. We think of the final cause of the object produced as anthropogenic because it is our *designs* that seem to initiate its production. The object so produced *appears* to us as finished when it achieves the material form that will fulfill its intended function. We are therefore drawn to say that this achievement is its telos, the final end to which its construction was always directed, and also the way it will appear at the completion of its limited task as having no further purpose: having helped Elon Musk ameliorate his insatiable desire for publicity, the car has completed its task and is now both driverless and purposeless, all the more so since, having escaped the earth's confines, it cannot actually be recycled as raw material for other ends.

Heidegger (1993, 290), then, suggests that ideas of causation, including the idea of a final cause, are not something "fallen from heaven as a truth as clear as daylight" but, in this case, something inextricably associated with that limited (technological) mode of engagement or addressing the world, a mode that also frames and limits our understanding of the world. As this mode of addressing the earth comes to predominate, so we become apt to consider all aspects of our existence, our being-in-the-world, in this terribly limited light. For example, the book that so engrosses us that the everyday world falls away as we read is reinscribed as just "print technology" useful in imparting information—or perhaps just a waste of more *productive* uses of our time. The love for a child is to be scientifically explained as a matter of *reproductive* efficiency, or just treated as a means for marketing toys. Heidegger's own examples famously include the river transformed into a source of hydroelectric power and the forest reduced to nothing but a source of timber. Such an orientation eventually becomes "second nature" to those under its thrall, even to the extent that referring to the whole earth as an assemblage of resources is assumed to be an ethically, epistemologically, and politically neutral description of reality! In other words, we come to (mis)take a specific technological mode of addressing the world as a fixed ontology of a now "objectified" world, and "having purpose" becomes entirely associated with involvement in telic productive activities.

This technological enframing of the globe is largely a product of the progressive imaginary, but it has a much longer heritage in Western philosophy. It does not yet entirely dominate the providential imaginary, but it is certainly present in its formulations and in much earlier accounts. Perhaps Western theological requirements for the universe to have a creator with a purpose in mind might themselves be indebted to this same mode of engagement: God as architect of the world; God as providentially organizing and making the world to have a certain purpose; God's *designs* on earth (in this sense, humanism is residually theological in posting humanity as a "whole" as having a God-like power to produce and progress toward "our" own ends).

Take, for example, Augustine's providential use of this technological trope:

> Let us not, then, deem God inferior to human workmen, who, in proportion to their skill, finish and perfect their works, small as well as great, by one and that same art. (as cited in Fergusson 2018, 14)

This, though, is a form of political theology that, to invoke Latour's earlier phrase, paralyzes nature, evolution, and the earth. It frames the earth's creativity in appallingly limited ways. This is also the reason Heidegger returns to pre-Socratic philosophies, and specifically to earlier notions of *physis,* to find alternative understandings in the Western tradition. As Sallis (2016, 52–53) argues, these earlier traditions offer a figure of nature that refers the questions of creation and origins back to the earth:

> Empedocles regards the *ἀρχή* [*arche*—that from which things arise] entirely in relation to generation, not the generation of being, but the generation by which vegetative life germinates in the earth and emerges into the light and the open air, as well as that by which animate beings are born. What is, perhaps, most conspicuous by contrast with later Greek philosophy is the total lack of reference to making or production *(ποίησις, τέχνη)*. Beings are not regarded as if they—even those that belong to nature—were made by imposition of form on shapeless material. One could say that in this sense Empedocles' thought remains closer to nature, keeps it apart from the paradigm of human artifice, of *τέχνη* [*technē*]. There is hardly a trace of the contention that will erupt later between these two sides . . . to say nothing of the dominance of *τέχνη* over nature that will later set in.

Physis, then, offers a nontechnological way of understanding earth's creativity and a way of distinguishing between nature and artifice that does *not* depend on making a simple ontological distinction between humanity and nature. The earth is not technological, and not everything humans do is a matter of artifice; that is, humans do not *have* to be technological in terms of their modes of addressing the earth, and they don't necessarily require final ends.

There are, of course, many philosophers (see, e.g., Foltz 1995; Zimmerman 1990) who have followed Heidegger in tracing the dangers inherent in the technological way of understanding existence which, with modernity, has come to enframe the entire way the world is "made" to appear. We might say that technology (in its Heideggerian sense) has come to pervade and dominate every aspect of our materio-semiotic imaginary, to frame everything (including, eventually, ourselves) as "standing reserve," without our becoming aware of its existential, experiential, and epistemic limitations. Why else is it that those seeking to defend the existence of anything at all are first of all asked, What *use* is it (Evernden 1999)? when providing an answer to such questions only serves to further enmesh the entity within the very imaginary that threatens its existence.

These Heideggerian critiques of technology tend to focus on the instrumental reduction of the earth and humanity to a potential *means,* that is, to standing reserve. However, Heidegger's analysis also allows us to realize that this technologically derived notion of telos *(of final ends)* is also universally overextended. Realizing this, there is no reason (apart, of course, from the overbearing weight and historical reality of its colonization of every aspect of life and every part of the globe!) why this very limited mode of addressing the world should actually be thought to apply to every aspect of human existence, still less to nature's activities, or to the earth as such. *In other words, the continual technological and philosophical production of final ends or purposes (which are never actually final) does not have to be the "be all and end all" of earthly existence.*

As we have seen, Kant's distinction, used in moving from providential to progressive explanations of the human condition, is entirely caught up with preserving the exceptional dignity of humanity as a final end. We might suggest, *to the contrary,* that our *exceptionality,* our supposedly unique association with generating or being considered the earth's final ends, comes to be justified through an unexplicated recourse to an entirely unwarranted extension of a historically and culturally specific (Eurocentric) misunderstanding of technology to the whole world, an occurrence that actually now leads that world to a very *undignified* "destiny." This seems close to Heidegger's own views, despite the fact that he regards his own work as still maintaining a space for human exceptionality.

Returning to Lovelock, we might now think about his distinction somewhat differently. He does often imply that the earth as a "living organism" *is* purposeful and that its being purposeful evolves with its self-creating and self-regulating capacity. This, however, is not at all in the sense of serving some ultimate purpose, a final end that it is "driven" to reach. He is certainly not describing Gaia as having any kind of final purpose in a Kantian sense, whether as an "end in itself" or as a means to a final end generated by creatures that have, in some historical circumstances, come to consider themselves exceptional. The earth's purpose does not lie in becoming "standing reserve" for human projects. Rather, we might say, Gaia is purposeful in the sense that the earth is *as near as we can ever get to* (but is not) something that is an "end in itself"—having spent the last four billion years moving through evolutionary (nonteleological) transformations that eventually, by accident and circumstance, included the creation of that same self-deluding creature. In this sense, the earth created the very possibility of its being understood purposefully, but the realization of this possibility was not a matter of planetary progress! It is not something the earth had "intended," something planned by external deity or earth goddess, or something predetermined or inevitable, and nor, contra Kant, will humanity's progressive self-realization constitute the earth's final end. Earth will almost certainly continue for billions of years after we have become extinct. Humans do not bestow purpose on the earth any more than we speak for it. We are not its materio-semiotic culmination but just one of myriad different evolutionary occurrences, albeit one we should, presumably, recognize and be extraordinarily grateful for! We are born of the Earth, of nature. "*Natura* comes from *nasci*, 'to be born, to originate. . . .' *Natura* means 'that which lets something originate from itself'" (Heidegger 1998, 183).

Through approaches like Gaia theory, human beings are fundamentally decentered. We do not constitute, in any sense (however far deferred), a final end, which also means that we cannot, by anthropic fiat, ever justify defining the earth and its evolving creations as just a means to our ends, a resource to fulfill any theologically providential or progressive project. This flatly contradicts that imaginary that, following Kant, regards the earth as here only to provide sustenance and sufficient challenges to enable us to determine and reach humanity's own final end(s). In any case. the emergence of the "Anthropocene" and global climate change demonstrates that this anthropogenic project is certainly not going to plan; indeed, it makes it obvious that there is no plan whatsoever!

Climate change now makes immediate and real that of which the imaginary of progress remains in denial—that *there is no final end for humanity other than extinction* and, furthermore, that the most likely immediate cause of this extinction will actually be the materio-semiotic hegemony of that same progressive imaginary that instigates anthropogenic climate change![1] The Kantian notion of a progressively realized "final end," as something opposed to, and different from, the "great system of the purposes of nature" (Kant, as cited in Zuckert 2007, 128), was fateful and fatal, being grounded on a faith in, and a desire to maintain and extend, human exceptionalism. If Kant was correct in arguing that there are no final causes in nature, then this is only because there are no final causes *at all* in any ultimate or overarching sense. Ecologically and evolutionarily, this makes perfect sense and is very clearly and explicitly a key assumption of Gaia.

1. Though, as Danowski and Viveiros de Castro (2017) remind us, we had better not ignore other anthropogenic possibilities, such as nuclear war.

A Purpose-Full World? Or How (Not) to Address the Earth

HAVING REJECTED THE IDEA that humans are exceptional in the sense that they can self-generate (or by God's grace provide) a *final* purpose for the earth's (or our own) existence, how should we think about purposes as such? Should we decide to follow Kant's requirement that we eventually excise all reference to purposes in our rational explanations of the *natural* world but, having denied exceptionalism, also extend this to human existence, and even beyond what we now call scientific explanation? Should we make purposelessness an essential aspect of our novel imaginary, our worldview and practices? This might collapse any distinction between our explanations of nature and of artifice, effectively universalizing that same mode of explanation Kant encouraged in the form of a naively mechanistic philosophy. This "universal explanatory ontology" might take as its motto "there is no such thing as a purpose" or "efficient causation explains everything."

Eradicating all notions of purpose would prove a challenge not only to humanism but to all those aspects of self-understanding and everyday language where we are accustomed to framing certain practices as purposeful—including, ironically, justifying science's own explanatory projects. We need to recognize that dissolving and subverting the sovereignty of final purposes is not the same as denying the emergence of purposes altogether. We can, it is true, see arguments that suggest that any notion of telos (or progress) is inapplicable to the earth's evolution (Nitecki 1988; see also the following discussion); however, there also seem good reasons for employing some notions of purpose even where biology is concerned (Allen, Bekoff, and Lauder 1998). Why would we think the squirrel's sequestration of seeds and nuts for the coming winter months is not purposeful so far as the squirrel is concerned? More importantly, why would we seek to universalize a mechanistic understanding that is the very epitome, and one of the drivers, of the same technological enframing of the world, albeit now in the form of a posthumanism? As such, it would be no less colonizing and offer little or no resistance to the reduction of everything to standing reserve at the service of ever multiplying projects. This route, Baudrillard (2007, 62) argues, leads to the technologically inspired disappearance of the human, another form of extinction at our own, technically mediated, hands.

Alternatively, we might come to recognize a plethora of context-dependent situations in which it may, or may not, seem appropriate to speak of purposes. Some aspects of those contexts would be social, some ecological, some largely explicable, some less so. Purposes could not be rigorously defined as true or final because God given or human produced, but this obviously does not eradicate the fact that many activities include a temporary orientation toward a context-dependent accomplishment, albeit with ongoing reverberations and consequences. This might be envisioned as a kind inessential and contextual patterning of what appears as purposefulness constantly shifting on many different temporal and spatial registers throughout the earth's evolution and ecologies, emerging in and through many different modes of being (human and nonhuman), senses, and encounters. We might imagine an anarchistic (lacking an overarching, unifying, or sovereign principle but by no means chaotic) worlding, where some patterns of apparent purposefulness may quickly vanish while others may persist over many generations, where some are culturally specific, some are not even species specific, each enabling limited, provisional, and *never* final ends to be reached, each involving things and practices that are *never* reduced simply to their being a means to an end.

There would certainly be lots of room for debate about what constitutes a purpose, its relation to intent, design, futurity, conscious or unconscious anticipation, instinct, function, accident, desire, determinism, and so on (Allen, Bekoff, and Lauder 1998), but also, and this is important, about the limitations of language itself, particularly languages so constitutively bound to a technological imaginary for expressing diverse "purposes." Here we might decide that, even though the earth lacks an ultimate purpose, it is still, nonetheless, intimately involved as an ineradicable part of the context of any and all attributions or denials of purpose. This is why, even if we say that the earth is as close as one can get to being an end in itself (see earlier), this does not mean that it is on the way to becoming or producing its own purpose in the sense of a final end. It means, rather, that the earth's multiplicitous creative activities afford the possibilities of purposeful activities. It means as well that the earth is that which, in its own modes of existence, forces us to leave open the question whether there are purposes in nature, purposes that exceed or are quite different from those defined technologically. *Earth in this sense may be full of purposes.*

This also means that any way of speaking of purposes is impossible to limit to the human realm. All kinds of relations between all kinds of beings might appear as purposeful in certain circumstances. It might, for example, seem appropriate, despite evolution being deemed nonteleological, and even if denied conscious intent, to say that the flash of the deer's white tail serves the purpose of alerting other deer to danger but by no means appropriate to say that the purpose of the spring grass is to feed the deer. The "explanatory universalist ontology" would see both explanations as misuses of a term without any ultimate meaning, as two conjunctions that should be explained in terms of more or less complicated sets or systems of efficient causes. We, though, might rather say that words (even those describing efficient causes) never describe reality fully or directly. They express something of a context-dependent situation conjoined with more or less appropriate uses of language found in other contexts. (Metaphor, as Lovelock remarks, is important in science too!) Appropriateness is always a matter of interpretation, never just ontology, and it is never determinate.

Let us call this approach an "ecologically entangled wor(l)ding" (although we might later want to refer to it as an aspect of a provisional ecology). Its motto might be "*the ontological status of the Earth is indeterminate*" or "all apparent ends are provisional" or, to take us back to MacFarlane's iceberg, "language is inadequate for expressing what the earth is." This, too, provides a challenge to humanism, but what diversity of world*s* might this respect, what different understandings of nature might transpire or be conserved, what affordances might it recognize in culturally diverse ways of dwelling? Would *these* worlds be so universally precarious?

Let us be clear: the difference between these approaches is not one of *simply* choosing to see the world in one way or another, as we might be asked to choose between cold reality and utopian daydreams. Nor is it just a matter of adherence to different understandings of ontology, although ecologically entangled wor(l)ding does suggest that any ontologies of which we can speak are always already modes of understanding and *never* just the bare, final "what there is" of the universe. In other words, it suggests that ontology is only ever *posed* as universal. It also recognizes the limits of language (especially modern language understood *technologically*) in its general inadequacy to capture nature or the earth, and the ways that language begins to lead us astray when we take it as a reified representation of the world's ontology. Ecologically entangled wor(l)ding does *not* say that it is impossible to describe the world technologically or in terms of final ends and/or efficient causes; rather, it says that we pretty much *have* come to describe it completely like this without any such decision having been made, and this severely restricts the way we can address and be addressed by the earth. The problems of climate change, pollution, eroded soils, mass extinction, and so on that denote the Anthropocene are a consequence (an intended or unintended product) of the materio-semiotic realization of this mode of addressing the world. The difference between these ways of answering, then, is not so much about what the earth *is* objectively (we could not possibly know that, because our ontological categories themselves rely on judgments appropriate to our materio-semiotic condition) as it is about objectivizing the earth; that is, it *concerns how we address the earth.*

This might, once again, be illustrated by returning to Lovelock's distinction. The very language of purpose becomes confusing here, where the difference seems to be whether or not something, planet or spaceship, might, or might not, be homeostatic, creative, living, an end in itself, a final end, and so on. An "explanatory universal ontology" might approach Lovelock's distinction by questioning its validity in terms of whether there really is any fundamental ontological difference between a spaceship and the living Earth; whether life can possibly add anything purposeful to dead matter; whether nature and artifice are actually fundamentally different; whether every living being is a self-regulating system, by asking how we should define the *self* in *self-regulating system*; whether machines or cybernetic systems might be developed that can self-generate their own "purposes" in ways identical in all important respects to the purposes we ascribe to our own activities; and so on ad infinitum.[1]

These are certainly valid philosophical/scientific questions, but

1. We can also see that this mechanistic approach does not present a challenge to the progressive imaginary but rather sets it to work on new tasks and projects. For example, we might be increasingly sidetracked into technologically driven debates about the possibilities of autopoietic machines or transhumanism, those forms of "posthumanism" that are, as Cary Wolfe (2010, xv) points out, better understood "as an *intensification* of humanism," a faith in the indefinite futurity (and profitability) of human-*induced* "progress" still framed as the irresistible replacement and movement away from a spent nature. Scientistic fantasies, of leaving our mortal bodies and even an anthropogenically polluted earth behind to *colonize* "empty space" (a novel, extraterrestrial Terra nullius) by Terra-forming lifeless planets not only threaten to repeat history as tragedy and farce; they are fatal distractions. The earth *is* Terra-formed, and this is surely Lovelock's point.

they ask about the spaceship and the earth in ways that assume that they are both just *objects* and that there may be an objective, determinate answer to these questions. That Lovelock, otherwise so obviously mechanistic, does not ask or answer these questions is one of the things that seems so contradictory about his approach, but then, as we have said, in many instances, he *does not* actually feel that the earth should be addressed as just an object. Objectification is always a consequence of a particular mode of addressing things. It relies on assuming that one can take an external view that does not shape the way the world appears; that is, it exemplifies exceptionalism in its making things available for "explanation." From the perspective of ecologically entangled worlding, we are not, however, short of explanations; rather, we are dangerously short of modes of addressing the earth as something other than an object, a resource, a machine, a system. We keep looking for a word or theory that will encapsulate what the earth *is,* but what matters is how we are addressed by and address the earth.

Perhaps, then, we are beginning to have an understanding of how *not* to address the earth, that is, technologically or as an instrumentally appropriated whole, and have also begun to explicate a possible source of this error, one that links specific social practices with specific philosophical theory. It is not, of course, a complete "explanation," it could never be so, but it seems plausible and appropriate to our situation. To not address the earth technologically requires, at a minimum, that we abandon the materio-semiotic imposition of that understanding of final ends that was only ever based on human exceptionalism and the hypostatization of a technologically derived understanding of causation as a universal and necessary feature of the world. That means that we cease to consider our own activities as ever having a final purpose and that we cease to assume or posit any kind of final purpose for creation, that we *should accept the idea of a world without final ends.* This is anathema to notions of theological providence and overarching progress. It requires different (plural and nonuniversal) imaginaries. It requires that we be open to other ways of addressing the earth.

This also requires that we be much more aware of the dangers in ascribing purpose or lack of purpose to any event. How easily, in our current imaginary, the view that something lacks purpose comes to mean that it has no value, except in relation to something that *is* deemed to have a purpose. How easily the declaration of a purpose comes to mean a final purpose, where that purpose is all that ultimately matters (materially and semiotically, and ethically).

Moreover, we see how damaging it is for any event to be circumscribed in terms of a self-contained productive process—from raw material via efficient cause and imposition of form to appearance of the final end. Where, we must ask, in the description of the modes of causation that define *technē* and its technological extrapolation, is there any mention of what is destroyed in the "process," what is wasted, what is surplus to requirements, of the cascading and never-ending aftereffects of the act of production, of the social, economic, ecological, and so on circumstances of its production? Is the destruction of the tree not also a cause of timber production? Is waste not always a product of making? Are some products not always surplus to requirements? Are some consequences of technology not always unplanned, unknown, undesired? "Ecologically entangled wor(l)ding" at least recognizes that no process is self-contained, that no end is final, that no one is in a position to conclusively declare something worthless, that all ends and purposes are provisional.

Environmentalism might almost be defined as being the only modern ethico-political movement that, since its inception, has focused on this telic leakage—on the consequences and so-called externalities of the productive process, the toxic residues, the emissions, extinctions, erosions, the repercussions of genetic modifications, and so on, and on. Of course, environmentalism's shallower versions merely suggest ways of keeping the predominant structure of productive processes going via reducing, re-*using*, or recycling, by more efficient use of resources, energy, and so on, that is, finding ways of internalizing externalities, bringing everything into a more expanded productivism, as in an "industrial ecology." Such approaches generally accept, rather than challenge, the dominant structures of the progressive imaginary. They may, at best, suggest certain exclusions, spaces that might operate as refuges from, or momentarily ameliorate, its worst excesses, but as climate change exemplifies, these refuges only grow more precarious as the whole world becomes subject to capitalism's own "global system of purposes"; that is, as we earlier suggested, capitalism and nature come to be regarded as an inseparable "system." This is not to say that reducing consumption, reusing, and recycling are entirely useless (quite the contrary, they exemplify usefulness!), but perhaps they are most creative precisely where they do *not* assist the continuation of the current imaginary, where they offer genuine alternatives to systematized production/consumption, where they actually resist the extension of resourcism.

Our *purpose*, then, in rejecting the finality of ends through a critique of their technological indebtedness is not to eschew the ascription of purpose per se, nor to universally expand its occurrence and find purposes implicated in every entity and event as a necessary feature of the world's ontology (as in theological providence). Nor do we intend to provide a different definition of purpose that might (or might not) prove more ecological. Rather, ecologically entangled wor(l)ding seeks to elaborate the ecological and cultural entanglements involved in any ascription of purpose, its materio-semiotic situatedness, its partiality (in its inevitable failure to capture the whole and to be impartial), and its telic leakage, in short, its *provisionality*. A theory of ecologically entangled worlding aspires to assist in de(com)posing the progressive imaginary, thereby opening possibilities for incipient imaginaries that are not technologically or anthropocentrically delimited. The world desperately needs such imaginaries, and Gaia, we suggest, might offer an inkling of one possible opening on these. However, unless Gaia can offer a way of addressing the earth that resists technological enframing, any imaginary associated with it will remain caught, rather like Lovelock, in its web.

Thinking the Earth Provisionally

THE PROGRESSIVE IMAGINARY addresses the earth technologically, demanding it provide answers to the questions "what is it?" and "what can be made of it?" Such questions have become second nature to most of us, yet they clearly indicate a change of emphasis and orientation from the questions posed by the earlier European imaginary of theological providence. Here the questions were "how is this earth, created by God, intended to provide for humanity?" and "how can we interpret the earth's activities in terms of God's caring providence?" Today the questions we need to ask are quite different again, namely, "how might we address the earth if not theologically or technologically?" and "does the earth care?" These questions categorically reject the idea that the earth is merely a means to fulfill an overarching eschatology or anthropocentric project and do not expect an answer in terms of a single definitive way of enframing the earth. Rather, these questions simply assume that the earth is the indeterminately creative and inescapable place of our abiding, that it is the ecological context of our entire existence as human beings, and that this is no less true for innumerable other interdependent beings.

The providential imaginary, of course, had no problem with describing the natural world as purposeful. It was mistaken in addressing the earth as a whole as having a singular purpose and in thinking that this purpose was to serve humanity, but nonetheless the world was purposeful in the very literal sense of being "full of purposes." Wherever the theologically informed naturalist looked, she saw signs of providence in the ways in which everything seemed so very well suited to perform its God-given purposes. Some, such as John Ray (1627–1705), early naturalist, dissenter, and member of the Royal Society, were willing to recognize that the creator's purposes and care extended to more than just human beings:

> It is the great design of Providence to maintain and continue every *Species*, I shall take notice of the great Care and abundant Provision that is made for the securing of this end. (Ray 1691, 86)

Ray's *The Wisdom of God as Manifested in the Works of the Creation* (1691) did indeed set out to detail these specific and abundant "adaptations" and provisions in all their diversity, all of which he deemed the "effects of a sagacious and provident nature" (87). Ray suggested that the physical and behavioral suitability of each *species* (a term he was responsible for introducing in its "biological" form; Wilkins 2009, 64), for its natural role, could only be explained as a result of design. It was not, he argued, just a matter of there being a natural "mechanism" (92) or nature being "*Plastick*" or "*Spermatick*" (92); rather, it evidenced existence of an all-powerful "architect," "engineer," and "Wise-Superintendent" (93).

In this sense, it was Ray, and not William Paley, who instigated the field that came to be known as natural theology (Paley, as cited in Armstrong 2000, 4). However, although Ray foreshadows Paley's famous metaphor of the watch with its finely tuned mechanisms found on the heath (Thomson 2005) and the implication that the watch, and consequently a still more intricately tuned nature, must be considered to have a designated purpose and designer, Ray's importance does not lie in any teleological argument for God's existence. Rather, Ray (1691, 131) exemplifies an early but *cautious* adoption of a technological metaphorics, one that is still wary of mechanistic allusions, speaking of such human inventions as a "few dead engines or movements" as compared to living beings "performing their own motions." For Ray, human inventions are poor additions to, or parodies of, God's creation that "cease presently to move so soon as the Spring is down" (131). Ray's purpose in emphasizing providence is rather to bring about a realization that the reason "Man [*sic*] ought not to admire himself, or seek his own Glory, is because he is a dependent Creature and has nothing but what he has received" (133–34). Ray also explicitly argues that the creativity of nature has purposes that are *not* just reducible to instrumental means for human ends:

> It is a generally received Opinion, that all this visible world was created for Man; that Man is the end of the Creation, as if there were no end of any other Creature, but some way or other to be serviceable to man. . . . Yet wise men nowadays think otherwise. Dr. More affirms, "That Creatures are made to enjoy themselves, as well as to serve us, and that it is a gros [*sic*] piece of Ignorance and Rusticity to think otherwise." And in another place [More says], "This comes only out of Pride and Ignorance or a haughty Presumption, because we are encouraged to believe, that in some sense, all things are made for Man, therefore to think that they are not at all made for themselves. But he that pronounceth this is ignorant of the Nature of Man, and the knowledge of things. For if a good man be merciful to his Beast, then surely, a good God is Bountiful and Benign, and takes pleasure in that all his Creatures enjoy themselves, that have life and sense, and are capable of enjoyment." For my part [says Ray], I cannot believe that all the things in the world were so made for Man, that they have no other use. (127–28)

In Ray's providential approach, Creation has not yet been reduced to the natural equivalent of a workshop or machine, and there is a heartfelt recognition that all earthly purposes are not simply subservient to human ends. One might even think that there is an incipient ecology here in terms of recognizing the diversity of sensible "enjoyments" of lives that do not serve humanity and of which humanity in general is unaware.

Generally speaking, and despite notable resistance (in Romanticism, for example), Ray's caution is lost in the transition between imaginaries as technics too evolve and become an increasingly dominant element of the progressive imaginary. Just a century later, William Hutton, philosophical author of *An Investigation into the Principles of Knowledge* (1794) and one of the founding figures of geology as a science, was fully utilizing the machine metaphor in his *Theory of the Earth* (1795). Hutton still retained a providential outlook, describing the earth as "'a system in which wisdom and benevolence conduct the endless order of a changing world'; 'what a comfort for man,' he added, 'for who that system was contrived, as the only living being who can perceive it'" (Hutton, as cited in Rudwick 2014, 70). However, as a deist rather than a theist, Hutton was content to think that God, having designed the system for humanity, had then left it running without further direct intervention. Hutton

> called the Earth a machine in allusion to the steam-engines that were such a spectacular feature of the Industrial Revolution. Steam-engines demonstrated the huge expansive power of heat, as one phase in a repeated cycle that could continue to operate indefinitely. This cyclic character, Hutton argued, was just what made the Earth a natural machine. (Rudwick 2014, 70)

Two centuries later, Lovelock would recognize how Hutton's "system of the habitable Earth" (Hutton, as cited in Rudwick 2014, 69) clearly prefigured Gaia scientifically, while denying, as we have seen, that the earth was any kind of "contrivance" or in any way contrived for human benefit. It is so very important that Lovelock, rather than emphasizing Hutton's earth-machine, chose instead to emphasize the metaphor of the earth as "living organism," indeed (as the title of the book in which Lovelock [2000, 15] declares Hutton a forerunner of Gaia makes explicit) as a planetary *body* that now requires a "planetary medicine." He chooses to de-emphasize Hutton's mechanistic frame, instead arguing that

> the waters of the Earth, as James Hutton saw long ago, are like the circulation system of an animal. Their ceaseless motions (together with the blowing of the wind) transfer essential nutrient elements from one part to another and carry away the waste products of metabolism. The rocks themselves are like our bones, both a solid strong support and a reservoir of mineral nutrients. (38)

Once again, however, Lovelock illustrates how difficult it is to describe the earth, this time metabolically, without implying purposes, just as it seems so difficult to refer to the circulation of blood without referring to its purposes in distributing oxygen and how easy it is, if, for example, we say that one of those purposes is removing "waste products," to fall into a technological framing, even of living bodies.

Notice, though, that if we refer to what the bloodstream *provides,* then less seems to hinge on the ascription of purposes. Presumably, no one would disagree with the contention that the blood provides oxygen to the body or, to return to our earlier example, that the deer's white tail provides a warning to its companions. More expansively, even if it cannot be described as its purpose, it does not seem at all controversial to say that the spring grass *provides* for the deer's sustenance. Provision here means that which appears and is afforded by a conjunction or pattern of events—it does not need to imply intent, an explainable causal chain, or a teleology. There is no necessary design or plan required to do this (although, of course, such occurrences can sometimes be intentionally enacted if circumstances allow). Rather, it is a way of speaking of an evolutionarily, ecologically, and only sometimes socially articulated conjunction of things/beings. Also notice that although provision may not be designed, it is not a matter of pure chance but of billions of years of earthly evolution. Provision is that which happens to afford possibilities for the continuation of certain temporary (provisional) patterns of existence, such as a human life, an ecological community, or a weather pattern. It is surely not controversial, then, to say that the earth provides for many different things in many different ways.

This is to say, if our notion of provision is not theological, but more a matter of ecological provid*a*nce (i.e., of what the earth provides ecologically), this shifts the grounds of our understanding. When we say that the earth *provides for* any being, we do not mean that the earth as a whole deliberately sets out to provide what that being requires of it, that this is the earth's intent or purpose. We mean instead that the existence of this specific being and the earth are so conjoined, so inseparable, so entangled, that the very substance of that being's existence is in the earth's provision. It *is* of the earth and *in part* it is the earth. Its existence, its form, habits, and life, all the affordances offered to it, are earthly. These affordances are not just random possibilities; they are ecologically co-constitutive, patterned, evolving together.[1] Each and every being, however transitory, is held in the earth's embrace, and each, however small, affects the future patterning of the earth's provision. Some of these ecological relations can be described as purposive—as the feeding barnacle stretches out and rhythmically beats its feathery cirri in shifting ocean waters. Some appear as caring—the owl provides for its young; the stickleback defends its watery "territory"; the otters play, coiling round and diving under each other—but these purposes are not final in any sense, and each is in its turn entangled with everything else, with sunlight, atmosphere, forest, the river's currents, the tidal rock pool.

It seems simple enough to imagine such specific occurrences together with a few of their earthly interconnections but more difficult to try to envisage the earth as a whole in this light; indeed, it is so difficult that it is tempting to say we should think only in terms of a concatenation of particular, more or less natural places or events with which we have some contact: the city park, the meadow, the tree on the street corner, the falling rain that precipitates experience. Certainly these encounters are a vital source of our understandings. Yet today we cannot *not* try to think the earth precisely because we desperately need to critically engage with, and offer alternatives to, earth's (mis)representation, its abusive enframing within the progressive imaginary, and because we need to re-cognize the importance of so many earthly patternings—the deep time of evolution, volcanism, continental drift, shifts in ecology, the atmospherics of global climate change, and so on. Yet how

1. Although, of course, as the end of the dinosaurs exemplifies, there are random events, such as the impact of extraterrestrial meteorites, that can have a huge effect on the direction and possibilities of evolution. Gaia is also so very dependent on the sun's energy; the earth is not a "closed system."

to address the earth in words when language, as we have seen, is so inadequate and when concepts can be so easily reified, that is, when we slip so easily from partial interpretations and metaphors into generalized abstractions and grant more effective reality to specific concepts than the earth's myriad manifestations?[2]

Certainly we need to avoid describing the earth as spaceship or machine or as the realization or embodiment of supernatural design. The earth is not a human birthright nor, as certain proponents of the Anthropocene might insist, something now anthropogenically produced. Not a resource. Not an object. Earth is not a system (even an autopoietic one) or a goddess, not a social "construct" or any kind of artifact. "It" is not even an organism, superorganism, or just another planetary body. "It" is not even an it, because this impersonal pronoun exemplifies a grammatical separation between persons and objects—but nor is the earth a person. We cannot finally determine what the earth *is*; we can only say, or seek, or point toward encounters, words, images, ideas, experiences, understandings, that express and enable different, *less productive*, but more creative modes of addressing and being addressed by an always evolving earth. We need expressions that expose some of the damaging limitations of providential, progressive/technological lexicons yet can also provisionally offer very different interpretations of our earthly involvements.

One possibility might be: the earth can be understood as that which lets ecologically entangled worldings, provisional ecologies, arise and appear from itself, world as a disclosure of the earth. As such, the earth constitutes emerging geological, ecological, and evolving patterns of being, now including human patternings. The reference to patterns here is intended to convey something of the sense of more or less coherent and coalesced, discernible and per-

2. A tendency identified by Evernden (1999, 54), using Whitehead's apt term the *fallacy of misplaced concreteness*. This fallacy is made appallingly evident in attempts to account for the earth in terms of monetary evaluations of "ecosystem services" (Sullivan 2010).

sistent, and similarly reoccurring arrangements or dispositions of bodies, events, things, fields, forces, and so on. Patterns in this sense are not designs or blueprints but active and evolving expressions. Earth might be interpreted as the immanent recurrent and successive generation and movement of these intricate patternings and existent interdependencies at vastly different temporal and spatial scales such that some seem, but are not, immutable, while others appear entirely ephemeral. These patterns are instantiated as a multiplicity of radically different but interconnected manifestations, such as the patterns of approaching storm clouds, the branching and rooting of the elm tree, the binary fission of bacteria, the earth's magnetic field, the meandering and eroding riverbank, the seasonal migration of birds, the changing constitution of your microbiome, breaking ocean waves, rotting logs, the vast array of beetles for which "God" (at least according to biologist J. B. S. Haldane)[3] seems to have an "inordinate fondness," and so on, and on.

Consider how complicated the relations in each of these, and innumerable other, patterning instantiations are, each mediated in what we are accustomed to think of as quite different registers—material, semiotic, phenomenal—but which are themselves all manners of expressive existence (Smith 2001). Some patterns are concentrated, some diffuse, but involvements can be traced between and touch them all in various ways.

How, for example, is the rotting log (de)composed? It was branching elm, holding up into air, sprouting leaves eaten by caterpillars, aphids, beetles and their larvae, sheltering nests of summer birds, now becoming rich soil from which seed and sapling may again germinate and grow. The tree was, by osmosis and transpiration, drawing up rainwater percolating through soil, water once raised as vapor to clouds from oceans and leaves' surfaces, then released

3. Haldane had purportedly been asked by a theologically inclined audience to say what his biological studies had revealed about God, although this oft-repeated "quotation" does not appear in any of Haldane's written works.

and dropping as liquid or wintery flakes from stormy sky, fallen rain filling and shifting the course of waters streaming down to sea. Birds, carrying seeds, bacteria, and spores, migrate back and forth across seas following patterns of the earth's magnetic field. Wood was created from leaves' sun-powered photosynthesis, transforming atmospheric carbon to carbohydrates and then to lignins. Other nutrients were translocated through soil to other trees, some of different species, via fungal mycorrhiza entering or sheathing and connecting plant roots. The elm was felled by storm's high winds after death consequent to infection by other fungal spores and bacteria introduced by scolytid beetles that leave intricate patterns of their feeding engraved under now peeling and shedding bark.[4] Fungi softening wood replete with springtails, millipedes, also insects with symbiotic microbiomes that digest the materials resulting from lignocellulose degradation—microbiomes like but so differently constituted than ours, only recently discovered but already themselves depleted and altered by fungus-derived antibiotics. Rotting wood attracts ants and more beetles, some to find provision and shelter, some directed there to die by parasitic fungi infecting insects' nerves and bodies, altering their behavior to suit fungal "purposes," wreathing insect bodies in shrouds of fungal hyphae, bursting from chitinous exoskeletons to release new spores. And on *and on.*

And this description only begins to elaborate what happens to compose and decompose almost every log on earth in the "normal" pattern of events, that is, without sudden flood or volcanic eruption, firestorm or timber extraction. There is no end to these patternings, tracings, touchings, involvements, expressions. Whether phenomenal, pheromonal, phylogenetic, physical, phytochemical, phenolic, phytophagous, philic, phenological, and only now, in such moments as this, philosophically extrapolated. There is no possibility of trac-

4. Dutch elm disease in Britain has, since the appearance of new varieties of this fungal pathogen in the 1960s, destroyed some thirty million elm trees, virtually the entire mature population.

ing all these partially interdependent and interwoven movements, no single path to follow to an end, no pattern that predominates forever. Rather, these patterns, as the word implies, appear to recur with different rhythms and temporalities, whether of lives, generations, seasons, or days, and in subtly varying ways in every such instance. To consider this as if it were the workings of a machine is, as Lovelock suggests, appallingly derogatory; the earth far surpasses any technological or human frame.

The earth "produces," but not at all like a workshop or production line; indeed, in an important sense, "it" does not *work* at all. This does not mean that the earth is broken or failing but that it is composed in manners that have no overall object or purpose but are (de)composed in its own multifarious gatherings. We might even say that the earth is the original and inhuman "inoperative community" (Nancy 1991), the differential sharing out and exposure of materio-semiotic patterns of existence (Smith 2010):

> The unity of a world is not one: it is made of a diversity, including disparity and opposition. It is made of it, which is to say that it is not added to it and does not reduce it. The unity of the world is nothing other than its diversity, and its diversity is, in turn, a diversity of worlds. . . . Its unity is the sharing out [*partage*] and the mutual exposure in this world of all its worlds. The sharing out of the world is the law of the world.
>
> The world does not have any other law, it is not submitted to any authority, it does not have any sovereign. . . . The world is not given. It is itself the gift. The world is its own creation (this is what "creation" means). (Nancy 2007, 109)

The earth naturally creates a world out of itself, a world composed of many worlds, that is to say, many modes of the earth's patterning manifestations, each comprising, enacted by, and exposed to differently patterned beings. The world appears, but not just as it appears to me or you. It appears in the very different material, semiotic, and phenomenal registers of all the different beings that also, in their shared exposure to others, together compose the world. A bark beetle's world is, and is not, "my" world; these worlds touch upon each other in many ways. They are both earthly, but they *world* so very differently. "My" world is, in any case, not *mine*; it is not my possession, it is what appears in relation to the earth that is exposed in and through my being. It is what and how I am given, what and how I am provisioned, what and for whom my exposure offers materio-semiotic affordances. The manner in which the beetle and I are mutually exposed, each to the other, our world-sharing, is not a matter of equivalence. The beetle might appear in my world only as a flattened body on the car windshield, as a photographic image in a book, or indirectly in the browning leaves of the elm tree or those patterns left exposed as bark falls away. I might hardly appear experientially to the beetle at all but still have meaningful and material effects upon its existence. This mutual exposure, even the nature of its being in any way sensed by me as mutual, is also delimited by the materio-semiotic imaginary I inhabit. In the progressive imaginary, the elm bark beetle becomes delimited as an economically and aesthetically destructive pest to be eradicated, controlled, and managed, although in one sense, it is only the messenger, not the message, the carrier, not the disease. The beetle has its own complex materio-semiotic relations, its own modes of expression. They are not economic or political in the same way that mine are, but they have economic and political effects. (So there is certainly a provisional ecological ethics and politics to be further elucidated here [Smith 2010].)[5] All of us being(s) are, from our inception, involved in shaping and delimiting the affordances

5. In a previous text, Smith (2010, 391–92) suggested that radical ecology could be understood as a "provisional ecological politics. Here 'provisional' would mean (a) offering an always revisable—*provisional*—understanding of (b) how ecological communities are both composed by and *provide for* the (more-than-just-human) denizens that inhabit them and (c) giving due recognition to the constantly changing conditions and modes (the eco-temporal provisionality) of such provision and (d) recognizing the provisionality of (ecological) ethics and politics *as such,* that is . . . politics as *constitutive* of, and providing possibilities for, enacting 'community' rather than something contained within and limited by (more or less permanent) *constitutional* political forms or principles."

of the earth's worlding, but we should remember that humans are not, as they appear to be in providential and progressive imaginaries (or even in Heidegger's writings—see later), the worlds over-*seers*.

All this diverse activity, patterning, sharing, and composure appears, in some sense, "coordinated," although this term may be taken to imply too much—for example, an entirely harmonious composition or even a "balance of nature" (something Latour certainly rejects where Gaia is concerned) rather than just a kind of inoperative ecological earth (Smith 2013). Those of a providential bent or seeking an explanation for everything might certainly interpret this "coordination" as evidence of a single underlying principle—a design setting out what is going to follow and not just an immanent expressive patterning. That, however, is clearly not our intent. Rather, the suggestion is that these expressive patterns play out together, roughly reenacting previous performances, but never with exactly the same cast and always lacking director, script, or playwright, even though they respond to certain recurring cues. Today, Latour suggests, what modernity had taken to be just the scenery is suddenly revealed as an integral constituent of the play. This play never ends but, dreamlike, shifts and disturbs story lines, eradicates or changes characters, settles and unsettles.

Yet we might still want to ask, how are such performances sustained, or "what holds all this together"? *Ecology,* of course, is one (relatively recent) word for this holding together of inordinately complicated patternings of composition–decomposition. *Evolution* is another. Both play key parts in Lovelock's Gaia. Here, though, we again need to note the temptation to reify and define these terms in ways that make them fit within, rather than challenge, the dominant progressive imaginary. Darwin's achievement, for example, is often described as his having provided the "mechanism" for otherwise materially inexplicable and variously interpreted ideas of evolution. For example, Gould (2002, 63) states that "Darwin developed the first testable and operational theory of evolution by locating all causality in the palpable mechanism of natural selection," acting upon individual organisms each varying slightly from others of their species, all in "competition" for limited resources to enable them to successfully reproduce.

Things are, however, not quite so simple. Darwin, like John Ray before him, was an experienced naturalist, deeply involved in, and concerned with, nature. Like Ray, he remained somewhat wary of explicit mechanistic metaphors; indeed, *The Origin of Species never* actually refers to natural selection as a mechanism (Ruse 2005, 285), although Darwin does once use the phrase "whole machinery of life" (as cited in Richards 2002, 534) and often speaks of nature's "productions" and "works" and "contrivances" in direct comparison to human works (see later). The image of the machine is also used by Darwin as a point of comparison with, and elucidation of, natural selection (Ruse 2005). Given the cultural dominance of mechanistic imagery, and it (somewhat *ironically*) being regarded as the only alternative to providential and teleological accounts, this mechanistic perspective is how Darwinism was, and often still is, received (though see Richards [2002] for an alternative account). In this sense, and in accordance with Kant's earlier suggestions (see previous discussion), Darwin provided an *explanation* that opened the door to removing notions of purpose from nature writ large.

However, the evolutionary "mechanism" itself easily becomes reified as a kind of transhistorical force, a shaping tool, such that evolution becomes imagined as a kind of natural evolutionary "process" or a productive force acting upon nature's "raw materials." Natural selection becomes reified as a quasi-mechanical force that occurs everywhere and in every time in the same way, iteratively producing a better fit between species and something "external" we often call the "environment." Here teleology can be reintroduced, sometimes, in its most damaging forms, envisaging evolution as a "progressive" teleological process, preparing the way for its processual culmination in what was often a racialized and gendered depiction of a competitively successful modern humanity—the tree of life crowned by the white European male.

For the most part, and despite his historical milieu, Darwin abjured such images. However, he certainly did want to retain a notion of overarching evolutionary progress and perfectibility in the sense that those individuals who reproduced successfully and preserved their "profitable variations" (Darwin 1884, 64) were deemed "better" than those who were "competitively" eliminated without leaving descendants. This progressivism, as Gould notes, was surely indicative of nineteenth-century European cultural biases. It was also, perhaps, a way of countering what Gould refers to as the loss of "solace" provided by what we have termed the earlier providential imaginary as it become further undercut by "mechanistic" rather than purposeful explanations for evolution. This removed any psychological comfort provided by an image of an earth designed by a caring God. Darwinism shocked *but also suited its social times* to the extent that its naturalistic approach obviated the requirement for supernatural design, and facilitated objectivism, but still tried to retain something of a godless teleology.

This tension is palpable in many passages in Darwin:

> She [nature] can act on every internal organ, on every shade of constitutional difference, on the whole machinery of life. Man selects only for his own good: Nature only for that of the being which she tends. . . . It may be that natural selection is daily and hourly scrutinizing, throughout the world, every variation, even the slightest; rejecting that which is bad, preserving and adding up all that is good; silently and insensibly working wherever and whenever opportunity offers, at the improvement of each organic being in relation to its organic and inorganic conditions of life. (Darwin, as cited in Richards 2002, 534)

Nature here is reified, selectively gendered, and anthropomorphized. She never ceases working but tenderly cares; she scrutinizes for quality control and preserves all that is good but ruthlessly eradicates the bad; she progressively improves all beings and herself.

In this and other passages, Darwin tries to provide an alternative for this, now compromised, desire for providential solace by emphasizing the grandeur of this evolutionary vision and trying to ameliorate the apparent harshness of an unending competitive struggle for survival. For example, in what Gould (2002, 137) describes as Darwin's "softest" of all statements, Darwin states,

> As selection works solely by and for the good of each being, all corporeal and mental endowments will tend to progress towards perfection.

This statement hovers between the providential and the progressive. It replaces one form of teleology with another and makes a universal evaluative (and incipiently moralistic) claim about the consequences of natural selection for individual beings, a claim that seems difficult, if not impossible, to justify when the *ecological* repercussions of "perfecting" bodies/behaviors—think carnivory, parasitism, and so on—are taken into account. What works for the "good" of *each* being, in the sense of having received previously evolved attributes that make certain activities more closely allied to particular environmental circumstances, does not necessarily make its life longer, better, easier, or experientially richer (all terms that of course carry their own evaluative assumptions) and certainly has no such implications for those other beings with which it engages, often quite the contrary. How, then, could the "good of each being" be universalized, and how is it assured? What on earth could the good of each being possibly mean here? Perhaps we should recognize that Darwin's soft statements actually exemplify a scientifically and ethically unconvincing attempt to argue that progressive ends might, just possibly, justify the evolutionary means—it is, perhaps, the closest Darwin comes to attempting to portray evolution as theodicy, to declare evolution *unintentionally* providential.[6]

6. The immediate intellectual source of Darwin's unintentional providentialism may be Adam Smith's notion of the "invisible hand." Gould (2002, 123) goes so far as to state that "the theory of natural selection lifts [Smith's] entire explanatory structure *virgo intacta,* and then applies the same causal scheme to nature. . . . Individual organisms engaged in the 'struggle for survival' act as the analog of firms in competition. Reproductive success becomes the analog of profit." What Gould does not

Darwin's explicit notion of evolutionary progress assumes that all that now exists is an improvement on what went before, that previous adaptations were somehow "imperfect," but according to what standard could this ever be measured, when all we have is comparative reproductive success at that time, in that place?[7] The clear implication here, though, is that humanity, with our "superior" mental endowments, is the epitome of evolutionary perfection (at least until something better evolves). However, as Ruse (1988, 120) argues, this thinking is questionable:

> Humans are not *disinterested observers* from afar. We are *products* of the [evolutionary] *process* and, although fortunately for our well-being, unfortunately for our understanding, we are necessarily *end products,* in that we are still around to ask questions. Also that we are necessarily *good (or if you like "good") products,* in that we are still around to ask questions. (our emphasis)

Ruse's point about defining "good" in terms of a recursive anthropic interpretation of evolutionary history is well taken, but his expressing this point in overtly technological terms also speaks volumes. Ruse (2005, 300), in something of a Kantian vein, regards this mech-

---

speak to, though, is that Smith's idea of the invisible hand is itself derived from a providential paradigm that argued that "God was at work in the market economy" (Oslington 2011, 63). This providential theology is largely ignored in contemporary economic discussions of Smith's invisible hand, but it makes little sense without it.

7. There are, of course, debates about "direction" in evolution and also about whether evolution might, despite mass extinctions and cataclysmic events, be understood, in general, as bringing about increasing levels of biological and ecological *complexity* over time. However, there is no simple nontautological way to associate complexity with progress. For example, complexity may actually increase vulnerability; tardigrades might well stand a better chance of continuing to reproduce than humans in a radically changing world. Gould, to be sure, makes an argument for recognizing certain directional (but nonteleological) changes in evolution, but his view of progress could hardly be stated more strongly: "Progress is a noxious, culturally embedded, untestable, nonoperational, intractable idea that must be replaced if we wish to understand the patterns of history" (Gould, as cited in Nitecki 1988, 319).

anistic metaphorics as something that "contributes to the epistemic excellence of the science" despite his own argument that it is the "artifact-metaphor at the heart of Darwinism" (Ruse 1988, 121) that wrongly pushes us "towards progress of a kind. It is of the very nature of artifacts that we humans try to improve them. . . . They have a progressive history—and this I suggest tips evolutionists, Darwinians especially, into progressionism" (121). Yet, if this is so, then this metaphor's contribution to epistemic excellence must surely be questionable!

Now, as Darwin also realized, this progressivism is only scientifically plausible if we presume an almost *providentially* stable, or only gradually changing, environment where adaptation can operate incrementally and continuously, each generation becoming "better" and "better" fitted to or integrated with *that* particular environment. This, perhaps, illuminates Darwin's tenacious retention of Charles Lyell's excessive geological gradualism in his own explanatory model, but it also effectively underplays the substantial evidence for previous mass extinctions and sudden, unpredictable, and dramatic environmental changes on all possible scales, some very local, some global. In other words, what we might call Darwin's "ameliorative perfectionism," his attempt to assuage the image of an entirely indifferent earth left by a "mechanistic" interpretation of a nature that lacks beneficent design, is another article of progressive faith now to be severely challenged by climate change and the earth's sixth mass extinction. We are now in a situation where so very many "perfectly" well-adapted species will find that their "progressive" improvements count for nothing when they no longer fit with increasingly chaotic and rapidly changing circumstances. Indeed, it might be argued that only the ecologically fortunate, the exceptionally hardy, and the highly adaptable will survive. The parallels with contemporary capitalism and its promotion of "flexibility" as the new work ethic should be obvious and are in no sense accidental (see Smith 2019).

Evolution, then, as a concept, still has to struggle to free itself from the patterns imposed by the progressive/technological imaginary, and this cannot be achieved by viewing evolution or, for that matter, ecology "mechanistically," for example, regarding variation as just the "raw material" of natural selection interpreted as a productive process. In any event, despite the triumphalism of the "modern synthesis" in biology from the mid- to late twentieth century, where, increasingly, only explanations couched in terms of strictly neo-Darwinian natural selection were deemed permissible, it has since become apparent that there is no single "mechanism" for evolution. This is something, as Clarke (2020) notes, that Margulis played a key role in arguing. There are, rather, several different and relatively distinct patterns of evolutionary change, each itself an expression of the many different ways in which patterns, organic and nonorganic, touch upon and involve each other destructively and creatively, continuously refiguring the legacies of the past upon encountering novel situations. There are many influences and ways of enacting evolution, through horizontal gene transfer, epigenetics, symbiosis, symbiogenesis, evolutionary developmental biology, and so on.

Such revisions to Darwinian evolution are, of course, complicated and contentious issues because, though they do not undermine the basic theme of natural selection, they challenge its absolute hegemony within evolutionary theory and reshape its understanding. They also offer vitally important opportunities to rethink the concepts and metaphors we use to understand evolution, to recognize these terms' cultural indebtedness and proffer (nonprogressive, nonmechanistic) alternatives. Evolution happens by more than one path, and its flows are interrupted by many different events and modes of change, pattern effects, and transmissions. It is composed of all manner of involvements between patternings, many of which are not well described by the term *competition* or by the focus on "individual organisms" who are, in every case, ecologically, natural historically, materially, and semiotically compr(om)ised. As Margulis (1997, 273) points out,

> of all the organisms on Earth today, only prokaryotes (bacteria) are [actually] individuals. All the other live beings . . . are metabolically complex communities of a multitude of tightly organized beings.

Evolution and ecology are not mechanisms or processes; they are terms associated with ways of trying to explain the recurring and changing patterning of the world and are terms that act as potential, though complicated, placeholders for the emergence of (nonprogressive, nonprovidential) materio-semiotic understandings of a much-more-than-human world.

Humans obviously both add to and, all too often, as in the case of the progressive imaginary, disrupt, dissolve, or forcibly reshape these patternings through the uncaring imposition of economics, politics, ethics, and sociocultural and aesthetic patterns, which then continue to become constitutive of our earthly situation. Some of these reshapings, disruptions, and additions can be momentarily and situationally liberating for some humans; however, the progressive imaginary instantiates and imposes a technological ordering that fundamentally renders its own role invisible in enframing the earth and changes some patterns so drastically as to limit and even destroy their provisional affordances. In all this, though, and despite this, we are still of the earth.

Here, then, we might offer another provisional attempt at delimiting the earth: the earth can be understood as the *gathering* together, the sheltering allowance, of all these patterns and expressions, however discrete or nebulous, obvious or hidden, as the generative gathering and provisional sustaining of the (de)composition, of the patternings it lets arise.

The Gathering Earth

THIS EARTHLY GATHERING can be thought ecologically and evolutionarily, but in thinking of it in this way, we must also attend to the thinking itself, to the manners in which thought, too, is patterned and gathered here. Otherwise, we will simply think that our thoughts reflect or represent the earth *as it is* and overlook the materio-semiotic and phenomenal patternings, natural *and* cultural, in which all thought arises. We will fall back into thinking earth as an object to be studied from above, rather than something of which *all* thought is also a part. There is not a single thought, however vague, scientific, original, crazy, or clichéd, that is not of (belonging to, emergent from, part of) the earth! There are various ways we might approach this understanding, for example, through Bateson's (1987) ecology of mind, through material engagement theory (e.g., Malafouris 2013; Graham 2020) or ideas of extended mind (e.g., Clarke 1997; Sutton 2020) where we think *with* the earth and its worldly manifestations, not over and against it, just as the mountaineer's thought is composed by and of the rock she climbs. For present purposes, though, Heidegger's work seems most pertinent because he articulates earth as an ongoing disclosure and gathering of myriad worldings that is both provisional and context dependent.

Earth Heidegger (1993, 351) characterizes as "the serving bearer, blossoming and fruiting, spreading out in rock and water, rising up into plant and animal." It is explicitly associated with the ancient

Greek notion of *physis* (see earlier discussion), with the sense of things arising out of themselves, pushing up into and becoming manifest in the phenomenological "world" experienced and appropriated by living beings, as seedlings emerging from a rotting log are sensed and grazed by deer. But, Heidegger reminds us, that which emerges to presence is never *fully* disclosed within the phenomenal world. It is only revealed under certain limited modes of appropriation, under certain aspects, in accordance with certain interests, expectations, sensoriums, histories, frames, predetermined interpretations, and so on. The deer senses the seedling as something to eat, but there is more to the seedling than this. In this sense, earth *as such,* the mysterious source of material manifestations, always holds something back; it conceals, and remains concealed, within itself.

Heidegger focuses on the particular possibilities for experiencing and composing an understanding of the world. In this sense, our thinking of earthly gathering is akin to, follows on from, but is subtly different from, Heidegger's. That said, like Heidegger, we can *provisionally* (the term is used by Heidegger too) delimit the "world as the manifestness of beings as such *as a whole*" (Heidegger 1995, 282, our emphasis). That is to say, Heidegger's world is not just the manifestation or phenomenal world of any particular being (entity) or experience of being (existence), for example, a beetle or its sensing the rough bark of a tree, but of the world understood as a composite gathering together of that which constitutes a whole in its manifestations, a whole in which human beings have a capacity to come to realize that they only participate, as mortals, for a while. Of course, when he speaks of understanding the world as a whole, Heidegger is not arguing that knowledge of it could ever be complete, that we can know everything about the world. For example, "we" cannot actually imagine all these myriad worlds, like the beetle's, because "life is a domain which possesses a wealth of openness with which the human world may have nothing to compare" (255). Nor can we fit these worlds together in some "objective" sense—although technology, in Heidegger's terms, is precisely the attempt to produce a world composed of such objects. Rather, his point is that only human beings can experience and think of the gathering itself, that is, of the world as a world, the world as such.

At issue here, though, is the manner in which, for Heidegger, the world *requires* humans' participation for earth's manifestations to be gathered together as a world. Conceptually, for Heidegger, the earth is "that which shelters in coming forth" (Heidegger, as cited in Haar 1993, 57); it is the hidden but continually manifesting ground, the "spontaneous arising" (Haar 1993, 57) that "informs," pushes up into the world in its appearing. "The world is founded on the Earth, the Earth thrusts up through the world" (Heidegger, as cited in Haar 1993, 59). However, as already noted, that appearing is always partial and shaped by the way that world is sensed, enacted, and understood (in the modern human case, this is now exemplified by a technological enframing where everything is forced to appear as resource). Earth, then, has a generative role in Heidegger, and the world that manifests is by no means just a human construct, but for Heidegger, earth does not itself have a capacity to form a world *as such* without the presence and contribution of human beings. That is to say, Heidegger (1977a, 221) argues, the world is only a world *as such*, for "Man" [*sic*] the "shepherd of being." This is why Heidegger (1995) refers to humanity as "world forming" and all other animals as "world poor." The appearance of world qua world is dependent on an opening or "lighting of Being" (Heidegger 1977a, 229) that is essentially only available to human beings.

How can this be, since animals obviously have experiential access to the world, the earth becomes manifest, appears, to them, too, even if not in the same ways, or even in the same senses? However, Heidegger claims, for animals, the world appears only as a collection of specific features that captivate that particular animal in terms of disinhibiting, letting loose, its instinctual behaviors, for example, the leaf captures the attention of the caterpillar, stimulating it to feed. Animals, Heidegger might say, are environmentally spellbound, caught up in a world constituted only by particular aspects of their disinhibiting environment; they cannot experience the world as such.

Seeing things *as* things, rather than just responding to them as stimuli, Heidegger argues, requires a capacity to be freed from instinctive captivation. Only then can one "stand within a manifestness of beings" as a being (Heidegger 1995, 248). Humans can be open to the world *as such* and be world forming only because they can let things be (not be captivated by them). In other words, to experience the openness of the world "as such" requires being able to experience something as more than a matter of passing interest or as being at one's service. Indeed, things show themselves *as such* precisely in their (in)difference to us, in the revelation to us of a withholding (as, for example, Heidegger [1995] famously argues, occurs in boredom) that also signals our participation in a world that exceeds our own singular existences. Insofar as human experience and/or thought facilitates an *exposure* to this withholding, it allows us to achieve a sense of the world as a whole, a world that exceeds our mode of access, a world we understand as existing even when we, as singular beings, no longer exist, have died. The awareness of the finitude of our existence exposes the world as such. Only "Man," according to Heidegger, has this possibility.

Now to regard animals as uniformly beset and enveloped in their instincts (whatever "instincts" might mean) is, as many, including Derrida (2008), have pointed out, asinine. There may also be, as Derrida further suggests, some doubt as to whether humans can, in actuality, *do that,* that is, think the world (ontologically) as a world as such or the things in it, the stone, the wind, the tree, *as such.* Of course, thinking the world as a whole, *for Heidegger,* is *not* thinking it in its totality; thinking it *as such* is not comprehending it as it "really is." Indeed, it is precisely the burgeoning awareness that beings, things, and the whole world have hidden generative depths, dimensions, and involvements that escapes our attention and resists our appropriation—earthly depths that language cannot fully encompass. Our understandings then, as Heidegger also indicates, can only be "provisional." To comport ourselves to the world as world, the tree as tree, is to recognize its dependency on earth's worlding *and* the withdrawal of its being in the very mode of its appearance.

The "as such" is the recognition of this unfathomable generative excess, the inevitable failure of any attempt to grasp everything about the world because of its dependence on the sheltering earth.

Still, insofar as the animal is generalized as "deprived" or world poor, the stone declared worldless, being world forming is cast as a privilege that sets an unbridgeable gap between humanity and all other earthly beings. Heidegger's approach is indubitably a form of humanism rather than, say, just a recognition of a multiplicity of specific differences. Evolution and ecology both resist this absolute and universal privileging (Smith 2017). We also need to recognize that, for Heidegger, the earth has no "natural history"; it manifests seasonal and sidereal cycles, but it is just generatively grounding—there is no sense of deep time or evolution, of the pre- or posthuman world or radically changing patterns of earth's modes of manifestation. Ecology too—the innumerable ways in which being-there is always dependent on and party to an ecologically entangled and changing patterning of earthly being are also muted, made to appear philosophically only in ways where they serve, resist, or affect our being-there.[1] Both evolution and ecology would be suspected by Heidegger of being just humanly degrading forms of "biologism." Yet, if we are drawn, as Heidegger suggests, in recognizing our own individual mortality, to an understanding of a world that exceeds, precedes, and survives our own world, then we surely have also to be drawn to recognize earth's existence before and after us, together with the far from impoverished or imperfect worlds, like those of beetles, that preexisted humanity and might well survive its demise. These worlds did not cease to exist or be gathered together evolutionarily or ecologically in various, never fully apparent ways, without human presence.

We might say, then, that consequent upon the anthropically involved gathering of world as an "inceptual" whole, there is also the

1. Hans Jonas makes a similar claim regarding Heidegger's refusal to engage with evolution and ecology, but his account of evolution is completely humanistic, teleological, and progressive in all the ways criticized here (Vogel 2018).

possibility, indeed the necessity, to think this other prior, ongoing, and future gathering—the earth without us, without, that is, what Heidegger refers to as a world thought *as such,* a world lacking the supposedly unique opening that humans afford. To do otherwise, to ignore this calling, would parallel that failing to think our individual mortality that Heidegger so disdains. We need to think the limit of humanity as such, that is, earth's gatherings without human beings, and also the evolutionary aftereffects of certain human imaginaries on an earth without ends. Of course, such an earthly gathering may not be one where attention is paid to the question of existence per se, but earth is that which, in "its" aimless evolutions, provided the very particular occasion of the there-being that raises this questioning. The diverse world(s) generated by this nonhuman earth may not be gathered together through thought or language (although this depends on one's definition of *language*! [see earlier]), but nonetheless, our very emergence is evidence of a continual gathering of "its" ever-changing evolving patternings over billions of years before we existed. We cannot claim to be philosophers, "lovers of wisdom," and ignore this!

If we can, as Heidegger suggests, *provisionally* delimit the world as a whole, then we can, to some extent, provisionally delimit the earth, too, in much the same way. Our being-here is always an earthly being-there that transcends in so many ways our experiential and conceptual modes of being. Think of the gap between the aspects gathered together in our paltry attempt to explicate the rotting log in its interconnectivity and the gathering achieved by the earth *without any assistance whatsoever from us.* We only have an inkling, an intuitive (but multiplicitously and materially informed) sense, of all the being, experiences, reactions, emergences, and flows between interpercolating patterns and the literally innumerable aspects of these overlooked, unknown, inexpressible materio-semiotic and phenomenal manifestations—the wildly patterned earth. Here there is no single overarching principle of creation, no design, no aim or intent. The gathering earth far surpasses the conceptual and material gatherings of humanity. Ecological wisdom would suggest that we should not mistake our world forming for a "higher" realization, nor for earth's Terra forming. This, too, might be regarded as the wisdom sheltered in Heidegger's resurrection of the term *physis,* which, we would argue, fits so very well with nonmechanistic notions of ecology and evolution.

Here, then, is another provisional delimitation of the earth, earth as *physis* beset by a specific and now dominant pattern of technological world forming comprising a progressive imaginary. Earth as continually worlding and withdrawing, differentially emergent as materio-semiotic manifestations, reproducing (but never identically) intricate patternings of being and beings. Earth as gathering beings together but not by hand, eye, or external order or under a singular principle but as anarchic (de)compositions of these evolving, differentiating patterns on immensely different scales. Earth as generative and expansive within a narrow envelope of soil, sea, and atmosphere, manifesting in different sensory, material, and semiotic registers; sometimes appearing as provisional or cruel, listless or frantic, crowded or deserted, inconsistently constant, as generously dancing leaves and screaming winds, surging icebergs, gently falling snow, ephemeral beings and geological temporalities, pulsing, rhythmic, but always a little offbeat, sheltering and releasing the sudden and unexpected, interweaving the differentially accessible, carrying within it billions of years of natural histories and still entirely encompassing us.

A Caring Earth?

WE WILL RETURN to this understanding of the earth as generative, patterning, providing, (de)composing, and gathering. However, we should note that the diversity of earth's patterning and the earlier critique of Darwin's *ameliorative* arguments might initially seem to make any case for a *caring* earth harder to sustain (though it in no way undercuts Darwin's celebration of the wonder to be found in considering the earth's ecology and evolution). Indeed, someone might say, is it not enough that the earth has done and does all this: that in all of "its" unfathomable creativity, it has given rise to, provided for, and still continues to provision the existence of so many diverse forms in all their interrelations? Should we really require the earth to care too? Perhaps we should just accept that there is no possibility of this being the case and hence no solace to be found in such a notion, that care is a form of relation exemplified and reaching its zenith in human interrelations, and that even here, it is something of a scarce "commodity" only manufactured in certain social circumstances?

We might also question whether this notion of a caring earth is even helpful to environmentalists. To be sure, an earth duly certified "caring" could at least avoid being mistaken for just an object or a resource. That said, we can also see that being caring in our current materio-semiotic imaginary provides no guarantee whatsoever of not being subjected to the most callous and destructive treatment.

Capitalism really doesn't care who or what it exploits, and it can ignore or even quite readily manipulate and profit from care, for example, by manipulating you to buy the right "green" detergent. It also demands that many, in the "service" industries and beyond, especially women, must continually provide emotional labor in the form of having to appear to care (Hochschild 1983). "Have a nice day!" Similarly, in so many respects, capitalism frames nature as something that serves to provide emotional respite from labor, a holiday destination, a green space for leisure, and so on, even as it is being exploited.

So caring is not enough, unless we can also trace paths to very different materio-semiotic imaginaries, and here the claim is that envisaging or experiencing the earth as caring (or as weird, wondrous, inspiring, magical, and so on) might actually be *inceptively* important, that is, engendering a provisional openness that potentially informs future manners of worldly appropriation (see later). Even just trying to think a caring earth allows us to see a stark tension between the earth's modes of patterning and gathering and capitalism's world-forming technological globalization (Nancy 2007). Thinking about why, from within modernity, it is so impossible to consider the earth as caring also allows us to see how far capitalism, as an instantiation of the progressive imaginary, has taken us away from patterns of materio-semiotic interactions grounded in, or at least informed by, care—indeed, how care, like the earth in all its manifestations, has become merely something to exploit.

As we previously recognized, in John Ray's historical, cultural, and ecological circumstances, nature's provision could only be framed in theological terms as purposeful, as a gift from a caring God, but this was clearly something that invoked both wonder and gratitude. It did not *just* provide a license to exploit and export an anthropocentric worldview. Care for more than human beings was deemed immanent in the natural world. Ray, though, as we have seen, was well aware that in this latter respect, he differed from "generally received Opinion" (Ray 1691, 127; see also earlier discussion), and this difference arose through a life spent trying to articulate (not yet "scientific") ways of addressing the more-than-just-human world, its plants, birds, airs, and waters. As progress rather than providence increasingly came to inform and dominate the Western imaginary, caring provision was replaced by a sense of earth as standing reserve for colonizing projects. Breaking from these historical legacies will be difficult and never complete, because both Providence and Progress are obviously globally formative; that is, they have now irredeemably affected the patterning of the earth, including the ways so many of us act and think, shape and are shaped by our materio-semiotic mode of existence.

In reimagining a provisional ecological alternative, we clearly need to avoid reverting to earlier forms of theological providentialism, which means not simply replacing God with a hypostasized Nature, despite some of us feeling a genuine and deep sense of awe concerning earth's immense creativity. Lovelock is again a useful reference point here. He offers no prayers to Gaia but says he understands those who do. He fully understands why a more respectful orientation and a sense of gratitude for the gifts earth bestows are entirely appropriate. He seems to regret that "progress" destroyed the idea of a quasi-providential earth and has made us forget how we are cared for and provisioned by Gaia, though he rightly sees no room for this understanding within Earth systems science or systems theory (but see Young 2020). He is, for example, often nostalgic for certain aspects of earlier modes of relating to nature, writing of our growing disconnection from nature when he states that the

> concept of Gaia or of the world of nature has never *appealed* to town-dwellers, except as entertainment. We lost contact with the Earth when our food and sustenance was no longer immediately and obviously dependent on the weather. (Lovelock 2009, 148, emphasis added)

Here, as so often in Gaian texts, the earth's indifference to our fate is portrayed as the mirror image (and a consequence) of the historically inflected indifference toward nature characterizing modern societies. Lovelock seems to suggest that we cannot *appeal* to nature, to earth, or to Gaia (note the *complete interchangeability* of the terms in Lovelock's own usage) because nature ceased to *appeal* to us when, at some historical time (as we became urbanized), we lost touch with the earth. It is this process that for Lovelock, paradigmatically, has had such "destabilizing" (if one dares use a term so intimately related to ideals of a *balance of nature*) climatic consequences.

This raises so many questions. Is this a quid pro quo? If we become indifferent to nature, then will nature express its indifference to us—and if so, how? With a virus, with a hurricane, or perhaps just in the way we fail to notice the natural wonders that appear before us? What "language" will Gaia use to express itself, and how should/could we *interpret* this natural expression given the materio-semiotic constraints and affordances of different imaginaries? Is it a matter of just letting you know that your fate is of no further concern (the earth, like a previously rejected lover, is now done with you too) or, alternatively, a matter of our now being awakened to the very impossibility of the idea that nature ever could have been concerned with us in any way whatsoever that could possibly have constituted "care"? As we have seen, Lovelock flits back and forth between these very different approaches.

This difficulty in expressing how the earth might (or might cease to) care, this communicative impasse, reveals both the inadequacy of our language *and* why this question is potentially generative for opening different approaches. Any solution we might find will have to express the important differences between providence, progress, and provid(a)nce. The typically modern answer would be to uphold a distinction between human subjects (capable of care) and inert or mindless objects (ontologically incapable of care) placing the earth, qua the planet, firmly in the latter category. A different but no less anthropocentric and humanist distinction might be made between human organisms (capable of care) and the earth as a (quasi) living organism, "a single physiological system" (Lovelock 2000, 11) or "superorganism" (Lovelock 2009, 133) that never had any capacity to "care" for its progeny, *except* though the accident of human existence. That is to say, as Hutton himself says, we *are* the earth's unique heart and mind. This, so far as we can tell, is close to Lovelock's current position. However, both these solutions obviously depend on a typically modern form of human exceptionalism, something that the very idea of Gaia should surely challenge.

The modern reluctance to refer to the earth as provisional and caring, rather than entirely indifferent, obviously circulates precisely because we consider care as something only exercised with intent and *purpose,* as exemplified by certain ethical forms of "subjective" or sentient human behavior, for example, in idealized parent–child relations or in the care we sometimes bestow on our pets or, for some, on their automobiles. We provide the pet or child with food, we polish and service the car, although we might not consider such services indicative of (ethical) care if it was largely self-serving, for example, increasing the vehicle's future financial value. Care, in this ethical sense, also needs to be other-directed with little or no concern for personal profit or return.[1] In Kantian terms, ethical care is a manner of addressing and providing for an Other as an end in itself, not as an instrumental means to our own ends. But of course, as we have already seen, the earth does not fit within a Kantian system; rather, it exemplifies the ways all provisions are only temporary and imperfect and never provide for us only as ends in ourselves, because nothing ever *is* an end in itself or, for that matter, simply a means to an end—everything is always ecologically complicated/implicated.

Such complications actually apply to human care too: it is always temporary (although in some instances, it might last a lifetime) and imperfect and always involves the relational interdependence of others' materiality, meanings, lives, and deaths, as varieties of feminist ethics of care exemplify (see Gilligan 1982; Larrabee 1993; Curtin 1996). Feminist care ethics also explicitly recognizes the

1. Though it might also be possible to identify more "self"-centered moralistic prescriptions in terms, for example, of Foucault's (1986) care of the self.

importance of caring as an emotional involvement in specific circumstances and the vital importance of feelings of being cared for as something quite distinct from, and not subsumed under, the abstract or "objective" apportioning of ethical standing, as, for example, in Kantian rights. It was a sense of "being cared for" that mattered to those seeking solace from ameliorative Darwinism. A feminist ethics of care is always already (ecologically/socially) entangled and contextual; it recognizes incompatible, indeed incommensurable, requirements that continually challenge and potentially dissolve any categorical imperatives. An ecological ethics, where provision for one often requires deprivation, even death, for other beings, and where "perpetual peace" (ecologically speaking) is never going to be an option, might well take something like this "form."

It is not accidental, then, that notions of Gaia/Nature/Earth as provisioning were vitally important in early developments in ecofeminism (e.g., Spretnak 1986).[2] Radical ecologists, such as Merchant (2006, xvii), have continued to envisage a feminist "ethic of *earthcare* that views both nature and people as real, live, active entities" appealing directly to those "weary of seeing nature as a vast machine that can be fixed by engineers and technicians." Just regarding the earth as provisional, lively, and manifesting care fundamentally challenges this machine metaphor. This is only one of the many ways that ecofeminism has tended to be more sensitive to the unspoken limits of dominant modern philosophical frameworks; as Sandilands (1999, 24) argues, ecofeminism genuinely attempts to come to terms with its "Western centric limitations . . . and the role of global capital."

Perhaps unsurprisingly, then, the question of care recurs constantly in ecofeminism. There are also explicit attempts to argue why a feminist ethics of care is a more appropriate ethical framework for environmentalism than other ethical theories or frame-

2. Although Spretnak's Gaia came directly from her engagement with ancient Greek mythology and spirituality rather than via Lovelock and Margulis.

works, such as deep ecology (e.g., Cheney 1987). That said, even here, the question of whether and how *the earth cares* often becomes sidelined as attention focuses on the gender implications of how to care for the earth. There are, for example, many intricate and crucially important discussions of the complexities and limitations of gendered metaphors of mothering and of the obvious implication that, in current circumstances, women are likely to be expected to take on the additional inequities of caring for (mothering) the earth (MacGregor 2006). However, whether and how the earth cares is rarely questioned, although it is certainly assumed that "she" does in those feminist discourses/practices willing to be overtly critical of the progressive imaginary, as, for example, in work focused on neo-pagan religious models of a Goddess who "is immanent in nature" (Starhawk 1999, 35). Unfortunately, many such approaches are liable to elicit claims from "progressive humanists" that they are outrageously anthropomorphic, wackily spiritual, politically retrograde, and hence potentially divisive for feminism and environmentalism as political movements. This is specifically alluded to by Spretnak (1986, 22) when she recalls being told how "talk of spiritual values and the feelings of reverence for Nature which had been prevalent in the [German] Green's first [electoral] campaigns" were squelched by progressivist arguments.[3]

Still, the question of how the earth cares and of spiritual values is obviously explicit in forms of ecofeminism emerging from, or respectfully engaging with, Indigenous traditions that have retained a sense of the earth's creative and caring capacities despite the ravages of colonialism and globalization. Many such accounts directly challenge modern Western presumptions of "environmental insentience" and of a human monopoly on ethics and care (e.g., Povinelli 1995, 516; see also Povinelli 2016). Cruikshank (2005, 3), for example, relates Athapascan and Tlingit oral traditions where

3. Here there was also the justifiable concern that a so-called religion of nature had played its own part in Nazism.

> glaciers take action and respond to their surroundings. They are sensitive to smells and they listen. They make moral judgements and they punish infractions. Some elders who know them well describe them as both animate (endowed with life) and as animating (giving life to) landscapes they inhabit.

The cosmologies of many Indigenous North American cultures are quite literally grounded in the ways that the earth cares and actively provides for human beings, but by no means just for human beings, and explicitly suggest that the only appropriate response is gratitude for these gifts. Kimmerer (2013, 107), for example, describes how an Onondaga school day

> begins not with the Pledge of Allegiance, but with the Thanksgiving Address, a river of words as old as the people themselves, known more accurately in the Onondaga language as The Words That Came Before All Else. This ancient order of protocol sets gratitude as the highest priority. The gratitude is directed straight to the ones who share their gifts with the world . . . "beginning where our feet first touch the earth, we send greetings and thanks to all members of the natural world."

The Thanksgiving Address provides a way of addressing the earth that ensures that it is neither articulated as an object nor simply reduced to a resource or consumer goods:

> Gratitude doesn't send you out shopping to find satisfaction: it comes as a gift rather than a commodity, subverting the foundation of the whole economy. (Kimmerer 2013, 111)

Again, from within the progressivist imaginary, any such talk of the earth's provisions constituting "care" and a "gift" is likely to be considered unjustifiably anthropomorphic, but why? If we recognize that the earth precedes us, then perhaps human gifts are only recognized as such because gift giving follows certain preexisting and more inclusive patterns of earthly provision. Certainly our very ability to care and our sentience are gifts from the earth, but so are very many carelike encounters with more-than-human aspects of the earth. If the earth, undeniably and in reality, provides (however temporarily) for all those who live, if the earth has *quite literally created and given us all our lives* and (for a time and to differing degrees) the affordances that can sustain them, and yet *expects nothing at all in return,* then is the earth, through these activities, not actually as near as one can possibly get to being an expression of both gifting and ethics? The earth is, after all, and literally, quite self-less in this respect!

Indeed, perhaps, the earth's care is much closer to being a gift than any human form of giving where, as Derrida (1992) suggests, there is always a suspicion of there being a "self-ish" (in both senses—greedy and indeterminately associated with a self) return, whether conscious, subconscious, or through diffuse affects that permeate and compromise the boundaries of that self (see also Marion 2002; Smith 2005; Manolopoulos 2009). And whereas gratitude in the Western tradition is something that when received might be thought to compromise the ethical selflessness of the giver, a Thanksgiving Address, or gratitude in general for nature's gifts, can in no way compromise the selfless activities of the earth. It can, however, radically challenge, change, and affect the ways we relate to/with the earth.[4]

That modernity balks at describing the earth as caring is indeed a reflection, indeed a key instance of, its technological framing of the world. Yet why, if it is not odd to say that the deer's tail provides warning to its companions, that the spring grass provides sustenance for deer, or even that the earth provides for us all, should it seem so strange to say that the earth cares for us? Of course, as these

4. This points to a quandary at the very heart of dominant paradigms of modern Western ethics: the insistence that ethics be formulated as a selfless provision that first requires the establishment of a self to subsequently count as acting selflessly. This was less true of certain forms of premodern ethics and is explicitly denied in Levinas's (1991) account of ethics as first philosophy, that is, as an other-oriented concern that emerges *prior* to the ontology of the subject, although, unfortunately, though perhaps not unexpectedly, Levinas's position is itself compromised by his spiritual humanism (see Smith 2012).

examples make plain, it doesn't *just* provide for *us* (humans), or even provide for *us* all, all of the time, in just the way we expect or might desire it too. It does not provide a never-ending "cornucopia" of delights, although that very term stems from experiences of earth's bounty through its harvested gifts, experiences from which many of us are now, as Lovelock suggests, so very far removed. However, as we have already noted, earth does provide our very existence, and of course, as mortal and ecologically unexceptional beings who partake in the earth, we, too, will eventually find ourselves providing for other beings, whether or not we care to do so.[5]

Interestingly, some forms of feminist "posthumanism" are also now making careful arguments for recognizing the animacy of the earth in ways that highlight its involvement in creating matters of concern, and its involvement in any ethics of caring. For example, Bellacasa (2017, 191–92) focuses on the care generated by and in certain agricultural relations to soil:

> Relational approaches to the cycles of soil life in themselves can be regarded as disruptions to productivist linear time. . . . Caring for soil communities involves making a speculative effort toward the acknowledgement that the (human) carer also depends upon the soil's capacity to "take care" of a number of processes [*sic*] that are vital to more than her existence. . . . Foodwebs are a good example to think about the vibrant ethicality in webs of interdependency, the a-subjective but necessary ethos of care circulating through these agencies that are taking care on one another's needs in more than human relations.

Although Bellacasa certainly focuses on soil relations involving the generation of human care, she here suggests that care is something that percolates through earth whether or not humans are present.

Taking all of this into account there does seem to be a deceptively simple way in which we might say the earth is care-full, just as we

5. As Plumwood (2013) points out, modern funeral rights are focused on delaying this provisional involvement so far as is technologically and financially possible.

previously suggested it was purpose-full. Specifically, when we look at the earth as being full of purposes, some of those purposes are clearly matters of caring for others. Humans are not the only beings that care for their young, they are not the only beings to direct their activities toward the provision of others or even to give their very lives to ensure such provision. Now this might only seem to provide a vicarious answer to the question of whether the earth cares by providing instances of care between specific organisms in certain times and places. There is also, as we have seen, an important difference between recognizing an earth full of varied purposes and thinking of the earth itself as purposeful in an overarching teleological sense. Something similar might be said for care. However, in both a Gaian and a provisional ecological approach, these organisms do not just live upon the surface of but actually, in part, *constitute* the earth, and so it seems, *to that extent,* that care in all its varied forms is also constitutive of the patterning of the earth, its ecology. In other words, it is simply not possible to say that the earth is necessarily, inherently, objectively, or wholly indifferent to every life or concern, just that it may not care "as a whole"—care is present only when and where care becomes manifest. The answer cannot be a simple yes, the earth does care, or no, the earth does not care, but a recognition that caring is one (of very many) registers in which the earth provisionally manifests or worlds.

The modern presumption is that the earth is uncaringly indifferent, but how can this be so if it is partly constituted by care and, albeit imperfectly and temporarily, provides so well for beings, including humans, that an entire imaginary was previously grounded on the idea that this provision could *only* be explained by reference to a caring God? It provides no kind of resolution simply to say that because the natural world sometimes appears as harmful, sometimes as caring, it must, in reality, be neither. Rather, the earth is never just harmful, caring, or indifferent. Indeed, the earth is never just *there,* like an object waiting to be explained. Rather, it is always creatively present in and as it continually manifests, as it worlds in all its complexity, a complexity of which we are now, and for a little while, an integral part. The ideas both of Gaia and of provisional ecology require that the earth be addressed not as something (an object) that stands over and against us (subjects) but as something that is our home and to which we are party. So if we say the earth cares, we are not just making a claim that nonhuman nature provides and cares for humans, as we might have a vision of an external God sitting back and resting while his well-designed and well-oiled machine works away providentially. We are speaking of how care appears and disappears in the more-than-just-human world. A provisional/providential ecology is not all about "us," but whether we recognize certain occurrences as manifesting care is a matter (and a semiotics) of how we address the earth and the more-than-human earth addresses us. It is a matter of interpretation *and* ontology together, not one or the other.

A reticence to regard the earth as caring also persists because we tend to think of care as something given continually, protectively, and irrevocably, whether directed at an individual or expressed as a more general attitude toward others. In other words, care has an indelibly ethical connotation, and so many of the things that happen "on earth" seem entirely incompatible with the notion of care or ethics. As John Stuart Mill (1885, 28–29) pointed out, nature hardly seems to be a perfect moral exemplar:

> In sober truth, nearly all the things which men are hanged or imprisoned for doing to one another are nature's everyday performances.... Nature impales men, breaks them as if on the wheel, casts them to be devoured by wild beasts, burns them to death, crushes them with stones . . . starves them with hunger, freezes them with cold.

Indeed it does. Many, perhaps most, aspects of the earth's manifestations cannot, by any stretch of the imagination, be deemed caring. But the earth is also present in everything that does care, in every instance of human and nonhuman love for others, in every act of kindness, every moment of wonder, every aspect of provision for every living thing. Of course, it can also appear, as Macfarlane's description of the iceberg demonstrates, as starkly indifferent, as weird and alienating, as terrifying, entrancing, nauseating, magical, and so on. Indeed, the earth appears in so very many ways at different times and places and to so very many different beings with immensely different interests and concerns.

Here again, though, these patterns are found in human care too; a caring relationship with another person does not mean that the carer always is, or for that matter ever appears as, caring to that person all the time and in every circumstance, still less that they care for everyone else in a similar manner. It does not mean that care for one may not have negative consequences for another or that caring is in any sense reciprocal. Care can be provided without love and thereby be almost indistinguishable from mere provision, or it can be intense and heartfelt and sustain and nurture in ways that comfort the recipient. We do not have to know we are giving care to provide it or to know that we receive care to be provided for, though both may be important. Of course, aspects of care may appear in human relations that rarely if ever appear in any other cases, but these are not consistent across all cases, and aspects of care might appear in other species that do not appear in human relations. Caring, again, is a matter of ontology and interpretation, which is not at all to say that we can interpret its presence or absence as we choose.

Not every provisional relation is going to be described as caring, but many that are do not necessarily require the constant intentional giving over of one's own concerns to serve those of another being. Think of the "ecological" complexities of how the provision of human care does and does not fit this pattern: the sleep-deprived parent continuing to begrudgingly and automatically respond to the child's insistent calls to be fed at all hours of the night; telling someone something that he *really* does not want to hear or keeping something hurtful from him; the hospital charged with an institutional failure of care despite the best intentions and exertions of its nursing staff; the last rites performed by a priest who has lost his faith; letting someone die or never letting go because we love her so very much; caring deeply who wins the election or last night's football game; lying about Santa Claus—or telling the truth; having no care in the world; caring for a toy; feeding a pet who regularly eats the local wildlife; caring whether your child passed her exams; caring what the neighbors think. Now think of the cases we might come across in a rotten log. The female wood louse keeping its young, so vulnerable to desiccation, in a specially created brood pouch under her shuffling body; ants attending to eggs and larvae, or perhaps to the aphids some species capture, feed, and "milk"; the logs becoming a tree "nursery" protecting the seedlings' roots and supplying nutrients; the mouse that protects then evicts each successive brood of its young every few weeks.

Care is circumstantial, relational, provisional—not something injected into a situation but an involvement of various patterns arising in the ecology of lives. There is nothing essential about care; it is not a singular kind of relation but a word we use in so very many different circumstances to try to express something of situations where we find overlapping patternings of provision that are not self-directed and concerned. Care appears as expressions and configurations of relational patternings of existence that for a while, whether intentionally or not, touch upon and serve to sustain or succor particular beings' and things' all too brief existence in the face of forces that would otherwise dissipate or damage them sooner rather than later.

How, then, are we to answer the question whether the earth cares? Should we try to solve this question like a difficult mathematical equation—the earth is, on average, 7.24 percent caring, 11.2 percent vengeful, 30 percent indifferent, and so on? Should we suggest a future dialectical synthesis to these unruly contradictions or perhaps return to thinking of the world in Manichean terms? Should we, perhaps, posit an ecological equivalent of a theodicy that seeks to transform every apparent evil into a prospective good? This, at least, seems initially possible, but surely only, as noted earlier, at the risk of hypostasizing Nature in God's place. It is also unnecessary. By presuming a perfectly good and caring God, theological providentialism *had* to generate a theodicy just as so-called natural theology required the supernatural to provide its ultimate meaning. A theodicy's purpose is to provide justification for appalling and unjustifiable happenings, often as an ultimate resolution of all suffering, in a final end that reveals God's designs and how, despite appearances, "he" really cares for us all (or maybe just for a privileged few believers), if not here and now, then at the completion of our earthly lives and/or the end of time. But this is precisely why this resolution is so *other*-worldly, because a provisional (temporary and nonteleological) earth could hardly play that role and so, at its theological best, is just temporarily accommodating rather than providentially directed.

If we decouple ecological provision from theological providence, we are no longer faced with trying to produce a *theodicy* of the earth, to prove that God's caring (but all so often hidden) intentions are best expressed in the earth's current ecological form. Nor need we be focused on "accentuating the positive" and "eliminating the negative" in our account of the earth's provisionality. We are not involved in some overarching Panglossian argument about this world being the best of all possible worlds nor engaged in the equivalent of a utilitarian calculus, weighing equable climate against dreadful storms, natural health against plague and pandemic (although, as Covid-19 demonstrates, these are so often "anthropogenically" and artificially enabled and transmitted, as unintended consequences of a technological enframing). We do not need to argue that all places on the earth are suited to human well-being or to teaching us valuable life lessons, nor even seek to belatedly justify some lost natural form of the earth as a golden age or paradisiacal place over and against modernity.

None of this means that our sustaining a sense of a caring earth is unimportant. If we recognize, experience, and respond to the ways that the earth provides *and* cares for us in fundamentally important ways, then we might address the earth very differently, resisting its being enframed theologically or technologically as something at our service. This relationship is not so simple as Lovelock's quid pro quo; the earth does not just become indifferent in a fit of pique if we are indifferent to it nor send down fire and brimstone like some vengeful god if we disrespect it. However, if we do, however unintentionally, address the earth as an object, as resource, property, commodity, or our birthright, this is indeed disrespectful to that which gave us birth, and it does indeed have materio-semiotic/ecological consequences, some of them potentially dire, certainly for many other Terrans, but for humans too. To regard the earth as, among other things, a source of provision and care, rather than a resource, might mean we stop leaving so very many other constituents of the earth to drown in our shit, although, of course, there are no final guarantees and no eternal reckoning. It might also provide some solace and hope against cold "realism" and/or alienated despair in the face of ecological disaster.

Attending to and Experiencing Earthly Provision and Care

> I was utterly alone with the sun and the earth. Lying down on the grass, I spoke in my soul to the earth, the sun, the air, and the distant sea far beyond sight. I thought of the earth's firmness—I felt it bear me up; through the grassy couch there came an influence as if I could feel the great earth speaking to me. I thought of the wandering air—its pureness, which is its beauty; the air touched me and gave me something of itself. I spoke to the sea: though so far, in my mind I saw it, green at the rim of the earth and blue in deeper ocean; I desired to have its strength, its mystery and glory. Then I addressed the sun, desiring the soul equivalent of his light and brilliance, his endurance and unwearied race. . . . I felt an emotion of the soul beyond all definition; prayer is a puny thing to it, and the word is a rude sign to the feeling, but I know no other. (Jefferies [1883] 1913, 4–5)

JEFFERIES'S ACCOUNT of his ecstatic youthful immersion in a world centered on the Iron Age hill fort of Liddington Castle in Wiltshire describes an earth that he felt actively provided for, responded to, and amplified his existence. This passage, which extends for several pages, begins his autobiographical *The Story of My Heart* and illustrates why, if Jefferies has a reputation today, it is, unsurprisingly, as something of a nature mystic.[1] He writes openly of an "intensity of

1. Jefferies was author of a postapocalyptic novel, *After London* (1885), in which nature has overrun every sign of "civilization"; of children's books where animals speak, such as *Wood Magic: A Fable* (1881); and of numer-

feeling which exalted me," an "intense communion" (6) held with the "strong earth, dear earth," (7) where an "inexpressible beauty of all filled me with a rapture, an ecstasy, an inflatus" (6–7). He speaks of "breathing full of existence" (14) and even "losing . . . my separateness of being" (10) while being able to sense the earth's deep time, "to feel the long-drawn life of the earth back into the dimmest past" (14). His sense of a "soul-nature" (13) that lies "far beyond my conception" (15–16) is incited by earth, sun, stars, but also by the natural historically imbued rural world of the nineteenth-century English countryside.

Jefferies does not confine his attention to the nonhuman world and is in no sense misanthropic, but it is here, keeping company with grass, birds, and sky that he experiences an intense sense of his involvement, of being "plunged deep in existence" (15). These profound experiences are initiated and continually reconfirmed through his attending to, touching, and being touched by specific instances of what we might term "ecstatic provision," for example, sunlight shining on the surface of his hand, humming bees, calling and leaping grasshoppers, or "flecks of clouds dissolving" (15). Jefferies's much loved world is constituted as he lies prostrate on Liddington's grassy slopes, delving into the chalky soil with his hands or attending closely to the lichen on the rough bark of the oak tree against which he leans. In these moments, the world appears both provisional and caring, resonating, fusing, with deeper patterning:

> I cannot leave it; I must stay under the old oak in the midst of the long grass, the luxury of the leaves, and the song of the very air. I seem as if I could feel all the glowing life the sunshine gives and the south wind calls to being. The endless grass, the endless leaves, the immense strength of the oak, expanding, the unalloyed joy of finch

---

ous popular articles on English country life, collected, for example, in the posthumously published *Life of the Fields* (1947). His most expressive book, albeit "a failure on publication" (Looker, as cited in Jefferies 1948, 139), remains *The Story of My Heart: My Autobiography* (1883).

> and blackbird; from all of them I receive a little. Each gives me something of the pure joy they gather for themselves. . . . The flowers with a thousand faces have collected the kisses of the morning. Feeling with them. I receive some, at least, of their fullness of life. (Jefferies 1947, 80)

Jefferies's experiences are so intense, excessive, and prolonged that they must appear almost impossibly different to any "normal" modern experience of nature. They do, however, resonate closely with Heidegger's explication of *physis* and the relations between Earth/World, even to the extent of a residual humanism where Jefferies ([1883] 1913, 13) claims "to have from all green things and from the sunlight the inner meaning which was not known to them."

The kinds of encounters Jefferies describes also fit so very well with Heidegger's (2012) explication of *Ereignis,* the *event* as a particular profound experience of the reappropriation of worldly existence. To partially translate *event* into our terms, we might say it is an opening of the epochal enclosure of our current (or any) materio-semiotic imaginary, revealing an inkling of the earth's transcendence beyond the enframed world of activities and appearances. Macfarlane's iceberg was, as already indicated, an event in this sense; Jefferies's experiences, too, have such affinities, though neither kind of experience is readily accessible to many of us today. That said, on occasion, something so very mundane as just standing open to the falling rain might suffice. After all, rain does not have to be understood, as vast arrays of waterproof consumer merchandise might suggest, as an encounter with falling objects we need to avoid at all costs, nor as something to be suffered, as by Samuel Johnson,[2] as a penance. Rather, given appropriate circumstances, rain, too, *happens* as a moment of inceptual worldly exposure.

In an earlier work, Smith (2017) suggests that rainfall can initiate a different mode of *appropriation*; it is an *event (Ereignis)* that

2. In his middle age, Johnson famously stood in the marketplace of Uttoxeter bareheaded and for a "considerable time . . . oblivious to the staring citizens and the pelting weather" (Wain 1974, 299) as a penance for his refusing to take charge of his father's bookstall there many years before.

has the potential to call us, however momentarily, out of technology and capital's narcissistic economy. Raining *happens* to allow things to come into view in ways not reducible to objects under the domination of a barren instrumentalism, not under the guise of commodities, or even under any predetermined theoretical *concept* but rather, as Heidegger says, "inceptively," where inception is "an enduring origin that opens up a whole realm of events and meanings" (Polt 2007, 382). The inception *(Anfang)* might be thought of as a cascading enactment of thoughtfulness rather than as the repetitive reproducible reification of concepts applied routinely to emergent contexts (Smith 2017, 232–33).

Exposed in our very being to the rains falling, outside of any productivist framing, we can, on occasion,

> gather something (glean an inception) of the elemental conditions of our fleeting existence here together, planted on the earth, watered by the vaulting, sometimes inclement, sky. That is to say, we can be momentarily exposed in our very being (verb) to the world's presence, its transience, and its transcendence—its becoming and its going on far beyond our sensory capacities [our words grasp] and our finite lives. Such exposure to the rains raining provides for our worldly existence in a proto-philosophical as well as a physical sense, not just as a thirst-quenching gift or even through its later suffusion and circulation, its collection and pooling, its flows and forceful springs, but also because it facilitates or calls forth an inception (a transitional phenomenological moment of openness that potentially informs all future worldly appropriations) concerning how we and things are gathered together, opened to the world's worlding. This is why being exposed to rain can be an event that reveals (and perhaps even allows us to revel in) the precarious and precious gift of life. (Smith 2017, 235)

The event is a revelation of the immanent concealment of the earth's excess, its withdrawal in the manifestations of its enframed worldings. Because of this, it also reveals something of the failure of any materio-semiotic imaginary to encompass, define, or enclose earth within the horizons of its modes of address. It can, we suggest, be an inceptive revelation of the earth as provisional ecology, as sheltered gathering, as *physis*. Such inceptional events are, of course, registered throughout ecological literature and many people's lives;

they can be profoundly life affecting, often deeply emotional, overwhelming, revelatory, but also simple, so very obvious, so thoroughly unexpected. Sometimes in such occasions, and subsequent to them, we are afforded a sense of being provided for, even being cared for, and also, perchance, a recognition of quite different instances of the earth's provision of care for other beings.

So many experiences can be dismissed as just occurrences rather than events, because they seem to leave no lasting impression on the patterns of our lives. They are not *inceptual* in Heidegger's terms and are easily brushed away. Their reception is also profoundly shaped by circumstances; their positive aspects, like a sense of being cared for, can be crushed by continuing histories of violence, by poverty, harsh words, ceaseless work, financial inducements. They can be shattered by suddenly shifting ground in earthquake, by plague or crop-destroying storm.

Importantly, and despite his blissful evocation of the landscape of his childhood, Jefferies also often speaks of experiencing the resistance that unfamiliar worldly encounters elicit, something, as we have seen, that Heidegger thinks important in allowing us to register something of the earth's hidden depths. In a passage that seems close to Macfarlane's description of the iceberg, there are times, Jefferies ([1883] 1913, 65) suggests, at least when "use has not habituated" our minds to their existence, when we encounter other life-forms in all their shocking otherness, as if "only at that moment they had come into existence." There is nothing familiar about them, no way to fit their appearance into the order of things, no understanding of their behaviors and relations; worse still, they appear coldly indifferent to human interests. Although not "inimical of intent" (65), their "anti-human character is at once apparent and stares at us with glassy eye" (65). Indeed, Jefferies states, a "great part, perhaps the whole, of nature and of the universe is distinctly anti-human," not set against humanity but "outre-human, in the sense of [being] beyond, outside, almost grotesque in its attitude" (64).

In such moments, Jefferies and the everyday human world are experientially decentered in the face of an inhuman excess. More than this, that which had constituted the familiar background is broken, shattered by these unfamiliar manifestations. For example, in describing a toad as "a shapeless shape appearing in an unexpected corner" (66), he notes that it

> sends a shock to the mind. The reason is its obviously anti-human character. All the *designerless,* formless chaos of *chance-directed matter, without idea or human plan,* squats there embodied in the pathway. (66, our emphasis)

Interestingly, the point of contrast here is quite explicitly technology (in Heidegger's sense) and teleology. The event, then, reveals the earth in very different ways, not only as a welcome source of provision, joy, or care but as something that explodes the current framework of our materio-semiotic imaginary, that patterning of self and social understanding and material involvements that acts as our second nature. This is not, as Macfarlane also shows, always a comfortable experience.

There is a stark contrast here. So many occurrences described by Jefferies seem to suggest that there is a sense in which the earth is patterned around us, speaks to and cares especially for us, and will always do so. That, after all, was also the promise of the providential imaginary. From this perspective, Jefferies's works might be read superficially as a kind of subjective wishful thinking, divorced from harsh reality, but this would be a serious mistake. Rather, as his life "progresses," Jefferies increasingly realizes that earthwise, we are drawn to the realization that we are just one of innumerable patterns provisionally sustained and then passing on and into others, and as he, and we, come to realize that this is *all* that nature offers, this can occasion anxiety, fear, repulsion. We might experience a horror exacerbated by precarity, old age, or illness that all too readily reveals that we are only ever a provisional locus of care, though if we are truly fortunate, care can be experienced in its ebbs and flows for as long as we live, even leaving traces, like wave patterns on beach sand, for a little while after we are gone.

However, when earthly provision and care appear finally withdrawn, the sense of abandonment can be terrible, as indeed it was for Jefferies. The ecstatic sense of exuberance in life can even, given different circumstances, descend into the kind of visceral disgust that the phenomenological psychology of Aurel Kolnai (2004, 73) argues is itself closely associated with "life-exuberance."[3] Here the

> surplus of life in disgusting formations signifies: accentuation, exaggerated representation, swollen overloadedness of vitality or of what is organic, as opposed to norm, direction, and plan of life, framework [*Gerüst*]. (Kolnai 2004, 72)

Kolnai's contrast between the organic and an anthropogenic technological form is evident here, although treated by him as timeless and universal rather than characteristic of the progressive imaginary. That said, Jefferies's references to grotesque inhuman encounters might evidence Kolnai's suggestion that, in certain circumstances, we are repelled by the excessive fecundity, the disorderly exposure, the fouling up or distortion of life itself and the loss of teleology and order. This can be seen in extremis in those common phobias of natural "objects" that express an all-consuming sense of deep-seated alienation from a natural world envisaged as refractive, invasive, amoral, indifferent, anarchic, chaotic, and intransigent, experientially initiated, both bodily and mentally, in the abyssal fear of the panic attack (named, of course, after the fathomless terror Pan, the therianthropic god of nature, could inspire in mortals) that dissolves any sense of safe self-enclosure. These phobias typically lead to a life patterned by the sufferer's obsessive desire to maintain order at all costs. They will compulsively clean and tidy their house; constantly police their bodily and material borders,

3. Kolnai's phenomenological investigations suggest that disgust can only, ever, be associated with encounters with living, or once-living, beings. "Disgust," says Kolnai (2004, 30), "is never related to inorganic or nonbiological matter." However, while "Kolnai's phenomenology helps link together the intentional objects and embodied sensations of disgust, it doesn't yet connect these to the larger social order" (as our notion of a materio-semiotic imaginary attempts to do) (Smith and Davidson 2006, 60).

checking for external invasion; seek to block all imagined avenues of contamination; radically alter behavior patterns to avoid any possible (or even quite impossible) encounters, taking only safe, familiar, and well-tested routes. Although given specific foci in spiders, snakes, thunderstorms, blood, and so on, each of which will induce extreme disgust and fathomless fear, these phobias are, perhaps, ultimately concerned with the way nature indifferently, unpredictably, inevitably, confounds all human-imposed order and all supposed progress by ensuring our inescapable mortality. The fear is perhaps indicative of the unavoidable realization that all our designs are to be interrupted as our own experiential world is inevitably eventually to be extinguished in its entirety and forever (Smith and Davidson 2006).

For all his worldly involvement and earthly intimations, Jefferies desperately desired that his experiences should transcend and escape nature, earth, even the cosmos. He had always interpreted the events in his life in terms of their offering to fulfill *his* desires to be party to the "strength," "mystery," and "glory" revealed in and behind the world's manifestations. These revelations, so beautifully described, became indicative of what he considered a higher meaning *beyond materiality* rather than anything more closely akin to Heidegger's earth. Jefferies struggled, especially as his life ebbed, to overcome the realization that his life as a whole might have no ultimate purpose. Certainly he could find no meaning in a purposeless nature, or in a deity that left the world to chance. Still, he imagined and tried to find words for something "more perfect" still, something "Illimitable" (Jefferies 1948, 259) that he sought through his own inadequate sense of his soul (134–35). Despite his earlier evocative descriptions of a "loss of separateness," he could never, it seems, allow himself to think that there was not an inseparable gulf between humanity and nature. Toward the end of his life, he wrote,

> I am separate altogether from these designerless things. The soul cannot be wrested down to them. The laws of nature are of no importance to it. I refuse to be bound to the laws of the tides, nor am

> I so bound. Though bodily swung round on this rotating globe, my mind always remains at the centre. No tidal law, no rotation, no gravitation, can control my thought. (Jefferies [1883] 1913, 69)

For all his astounding insights, his careful attention to other beings, the revelatory events that opened and patterned his life work, he always, sometimes joyously, sometimes desperately, struggled to think his own existence as something more than just "ecological"; he did not want to believe that he, too, would be carried out by the tide.

In his final days, he wrote,

> Mistake—absurd veneration of works of nature as if they were divine—no mind in it at all. How marvellous! Wrong. Porcelain more wonderful—clock—nature has been working for 250,000 [*sic*] years and without any drawback of pain or nerves. If a man's mind had been working with irresistible forces at beck and call for 250,000 years he could have done something better than this. No Great Beyond then in this—the Material or Matter. Nature works by mistakes and failures. Nature is very stupid. Innumerable mistakes. Appears to do nothing but make mistakes and so ultimately arrive at something after continual blotting. Our intelligence seems to come to us in some way from outside nature (or matter). (Jefferies 1948, 260, written May 6, 1887)

This reaction has to be understood in a context where nature seemed, at that moment, inaccessible to him (he literally could no longer walk in nature but was confined to his room) and to have prematurely dashed all his future designs (he was only thirty-eight), refusing him any care or consolation. The life-threatening disorder of Jefferies's own guts as he wrote *The Story of My Heart,* the hellish and putrescent associations of the "intestinal tuberculosis" (Williamson, as cited in Jefferies 1934, 13) and anal fistula that he knew were leading inexorably to his own "premature" death, meant that body and mind proved increasingly unable to afford any possibility of "prayer" with the earth or allow him to dissociate morbidity from life. He sought salvation in the kind of supernatural musings that he thought nature could never offer. His last diary entries switch back and forth, from thoughts momentarily landing on plants, the flight of birds, shells, to his desperate situation: “Stitchwort. Beauty suggests hope” (May 1, 1887, in Jefferies 1948, 260).

> Swift. No hope, no gratitude, to whom! No love to turn to, yet the mind is not satisfied. How am I a [*wretched* struck out] poor short creature, teeth, mammae, nails, relics of brute time—to see into and define the Great Beyond? Curves of limpet and scallop. The whole earth deceives and throws the mind aside from true contemplation. (Jefferies 1948, 258, written May 2, 1887)

Jefferies so desperately wanted, as we almost all do, to hang on to his self-identity, his experiences, the great gift of his life, to “wander and sail forever. No limit” (289), but is palpably torn by the fact that any such possibility requires, in a cruel parody of Peter’s denial of Christ, the *denial* before night falls of any gratitude for its earthly sources. He recognizes that he is “out of tune” (Jefferies 1948, 289) with nature because of his illness but cannot resist placing the ultimate blame for this illness, despite all his experiences to the contrary, upon nature’s cold indifference:

> Nature is like a beautiful statue. I must love, must gaze. Yet I cannot put the life into it I should like to. Cannot make it love me, or do as I should like to see it. Still cold, however lovely. The sea is the sea and will not love you again. (Jefferies 1948, 289, written June 1887)

> I hate nature. I turn my back on it. *Works* of man greater than nature—Nature works without a mind as the sea sculpturing the cliffs. (Jefferies 1948, 259, written June 1887)

Jefferies died on August 14, 1887, at his house “Sea View” in Worthing, Sussex, his body buried in the same cemetery as that other great nature writer, W. H. Hudson. In a sense, his life exemplifies the experiential possibilities and the difficulties in believing that the earth cares. In so many instances, he obviously experienced a deeply relational world infused with beauty, care, and joy. However, if our measure of care is granting eternal life, limitless knowledge, endless progress, or complete spiritual enlightenment, then this is to apply an entirely artificial framework to an earth that can promise no such thing. Earthly care, life, and knowledge are always provisional and contextually entangled.

> Against this his ill-health is nothing to record, except as something triumphed over by the spirit of life. His sadness came of his appetite for joy, which was in excess of the twenty-four hours day and the possible threescore years and ten. By this excess, resembling the excess of the oak scattering its doomed acorns and the sun parching what it has fostered, he is at one with Nature and the forces of life, and at the same time by his creative power he rescues something of what they are whirling down to oblivion and the open sea, and makes of it a rich garden, high-walled against them. (Thomas 1909, 324)

At least for a little while, for all walls crumble, all gardens overgrow.

Provisional Ecology

IF MACFARLANE'S EXPERIENCE of the iceberg reveals something of the earth's inhuman manifestations (and Jefferies's "designerless things"), he also recognizes the importance of context in affording the sometimes troubling, sometimes exhilarating patterns of our earthly relations. Macfarlane is especially well attuned to the ways we (mis)interpret the phenomenology of nature's appearances due to our limited positionality, our egocentric concerns, and our cultural biases. He also constantly reminds us that life is not all about us.

In *Underland,* Macfarlane (2019) describes a perilous winter climb over the mountain ridge running down the center of the Norwegian island of Moskenes in the Lofoten archipelago. His long trek uphill, to visit the sea cave walls at Kollhellaran and the red dancing figures painted there some twenty-five hundred years ago, had become threatening, with the constant "rumble of small avalanches" (260). He experiences "a strong sense of the terrain's disinterest, which I might at other times experience as exhilaration but here, now, can feel only as menace" (260–61). Suddenly he falls up to his armpits into a fissure in the snow, "legs dangling in a void," only extricating himself with intense difficulty, "as if getting out of quicksand" (261)—eventually hauling himself up, he manages to climb "onto the saddle of the pass with a whoop" (263).

> I lie on my back, a gaffed fish, breathing heavily, and there above me, showing through the mist, is a sea eagle. Low and circling, and the

> queasy fear in my throat is forgotten and my heart leaps to be over-flown by that remarkable bird in that remarkable place. Then I think, it's just sizing me up as lunch—and I laugh out loud at my stupidity and the land's indifference. (263)

Granite and ice, cliff and crevasse, once again exemplify Macfarlane's remark about the indifference he often experienced in the mountains, their lack of concern whether he lived or died and how this indifference may even provide a sense of exhilaration, especially when danger has passed and the climb is accomplished. The appearance of the sea eagle initially seems a wondrous recognition of his struggle and survival at this pivotal moment. He is elated but then, almost immediately, *brought down to earth* through the realization that, from the eagle's perspective, his success just brings about a rather disappointing end: the loss of a potential meal.

Would it make sense to try to describe this event without recourse to a language of purpose? It could, perhaps, be done, but it would be both an uphill struggle and a thankless task. We are certainly forced to recognize that this is a landscape that dwarfs and is impervious to Macfarlane's own purposes. What purposes are gathered together here conflict, though both relate to survival in what might seem, at least to a human being, but perhaps not to an eagle, a hostile environment where Macfarlane is left feeling like a fish forcefully dragged out of water. Both human and eagle are concerned with the other, but not in any ethical sense. How weird it is that the mundane purposes of one constituent give rise to a momentary elation in another, that indifference might seem to be everywhere, but actually there is no sense that either the eagle or Macfarlane were in any way indifferent to the presence of each other. Isn't it also strange that being suddenly and unexpectedly *reconciled* to this can generate laughter born through puncturing the presumptions and pretentions of any ego- and/or anthropocentrism? Isn't there, at least provisionally, a sign of hope here? Might such events even appear mundanely providential, in the sense of serving to open us to the possibility of joy even in a place of harsh realities, a place seemingly void of care, but inhabited by beings that are only presently there because of previous patternings of care?

Such events can be inceptual. Are we here afforded an opportunity for imagining the world as ecologically provisional, the earth as providing, both indifferently and caringly, for the more than just human? Perhaps it will not even seem so weird to repurpose John Ray's (1691, 127–28) statement that the earth sometimes "takes pleasure in that all . . . Creatures enjoy themselves, that have life and sense, and are capable of enjoyment" and that it is no longer plausible to think that "all the things in the world were so made for Man."

With climate change, the exceptionalist ground of the progressive imaginary has begun to tremble; a different kind of earth-quake is coming. There is a sense that we are coming to the end (whether conclusive or inconclusive) of this particularly destructive and colonizing imaginary. The very "success" (let us rather say the globalized enactment and enframing) of the progressive imaginary ironically ensures that it loses its purpose, its obviousness, and its sense of direction. The accelerated consequences of the technological reduction of everything to standing reserve and its commodification are precisely what now undermine the possibility of thinking of *anything* as either a fixed or final end. Events, everywhere, every day, open up a realization that the progressive imaginary entirely fails to offer any final purposes, any ground on which to stand, or any comforting destination; it just produces increasing quantities of consumer products with short shelf lives, rapidly passing fads and fashions, debts, profits and discarded theories, mountains of personal data to be mined and processed, and so on. Capitalism and technology have, in effect, produced a totalizing and failing parody of nature, where we are economically, digitally, and chemically (de) composed by what we consume, where computer algorithms now decide the expendability of lives. We have to face it: "The end itself has disappeared" (Baudrillard 2007, 70).

What, though, if anything are we to make of events coming together and being conjoined to afford the possibility of such moments of realization? Perhaps nothing. But then again, perhaps this is a question not of us "making" anything but rather of trying to conserve (because preservation is never an option) something of others' threatened existences, of us *coming to care*.

Is this opportunity providential? Will it save us? Who could tell? Who cares? With theological providence, we always had to find the message elsewhere, to interpret God's intent, and we never dispelled final ends or ceased to interpret everything in terms of benefits to individual human beings or a larger human collectivity. The value of every earthly affordance was in the intent behind it, in the presumed relationship between God and humanity. Nature, the earth, was just the intermediary. With ecological provision, it is very different. There is rarely a message in terms of there being a specific intention or an intended recipient, and even where there are intentions or purposes, there is so much scope for misinterpreting them, as the sea eagle reveals (at least to the ecologically minded like Macfarlane). Some messages are received by us and interpreted in one way, some by another being and interpreted quite differently. Any event might provide a plethora of materio-semiotic affordances for innumerable and incredibly diverse fellow Terrans, and the "messages" are not always good! Some of these messages might take the form of an iceberg or an eagle, some of an argument.

Things might be experienced or interpreted as providential after the event, but nontheological providence would really just mean an acceptance that the event was inceptual, that it provisionally changed the patterning of the earth in some significant way. We no longer have to think of such events as providential opportunities sent by external forces to change our lives, or as moments of enlightenment, but they might, nonetheless, be provisional, affording openings into the evolution of provisional ecological imaginaries, generating care that would otherwise be absent. Such epochal events also allow us to recognize that even here, in the so-called Anthropocene, the earth may have been overwritten, but it has not ceased to exist and to (de)compose wor(l)ds.

Does the earth care? At issue here is the earth and speaking of the earth, and what the earth means beyond signification. The earth does not care as a whole qua a totality or unitary entity, but then we need to stop thinking of earth in this way. Rather, as an inescapable involvement, an anarchic gathering of differently changing patterns, innumerable ways of creating and holding things together, composing, decomposing, recomposing, as that which lets something originate from itself, care certainly is among the earth's worldly provisions. When we are afforded the possibility of letting the earth world differently (provisionally and inoperatively), care wells up as the spring after heavy rain, mysterious in its source but incontrovertible. We will never be the sole or privileged vessel or end of the earth's care. Wanting, we sometimes find care, or care sometimes finds us wanting. Care appears; it cannot be made, shaped, or completed, and certainly not manufactured. It does not have a final purpose; rather, *we are sometimes called, ecologically, by events beyond our control to provisionally become caring* in our earthly inclusion and our worldly exposure to others.

Outside the window, the snow begins to fall again.

Acknowledgments

We extend our gratitude to everyone in the Extinction reading group at Queen's University for their many generative discussions and insights, with a special thanks to Josh Livingstone for his detailed comments on the sections concerning Kant. Thanks as well to David Doyle for reading the entire draft and to Bruce Clarke for his generous review of the manuscript. Finally, we would like to acknowledge the innumerable and all too often overlooked more-than-human affordances that made the creation of this work possible.

Bibliography

Allen, Amy. 2016. *The End of Progress: Decolonizing the Normative Foundations of Critical Theory*. New York: Columbia University Press.

Allen, Colin, Mark Bekoff, and George Lauder, eds. 1998. *Nature's Purposes: Analyses of Function and Design in Biology*. Cambridge, Mass.: MIT Press.

Anderson, Benedict. 1991. *Imagined Communities: Reflections on the Origins and Spread of Nationalism*. London: Verso.

Apel, Karl-Otto. 1997. "Kant's 'Toward Perpetual Peace' as Historical Prognosis from the Point of View of Moral Duty." In *Perpetual Peace: Essays on Kant's Cosmopolitan Ideal*, edited by James Bonham and Matthias Lutz-Bachmann, 79–110. Cambridge, Mass.: MIT Press.

Armstrong, Patrick. 2000. *The English Parson-Naturalist: A Companionship between Science and Religion*. Leominster, Mass.: Gracewing.

Bateson, Gregory. 1987. *Steps to an Ecology of Mind: Collected Essays in Anthropology, Psychiatry, Evolution, and Epistemology*. Northvale, N.J.: Jason Aronson.

Baudrillard, Jean. 2007. *Why Hasn't Everything Already Disappeared?* London: Seagull Books.

Bellacasa, Maria Puig de la. 2017. *Matters of Care: Speculative Ethics in More than Human Worlds*. Minneapolis: University of Minnesota Press.

Benjamin, Walter. 1991. "On Language as Such and on the Language of Man." In *Selected Writings*, vol. 1, *1913–1926*, 62–74. Cambridge, Mass.: Harvard University Press.

Beradi, Franco "Bifo." 2011. *After the Future*. Edinburgh: AK Press.

Burns, William E. 2002. *An Age of Wonders: Prodigies, Politics and Providence in England 1657–1727*. Manchester, U.K.: Manchester University Press.

Castoriadis, Cornelius. 1998. *The Imaginary Institution of Society*. Cambridge, Mass.: MIT Press.

Cheney, Jim. 1987. "Eco-feminism and Deep Ecology." *Environmental Ethics* 9 (2): 115–45.
Clarke, Andy. 1997 *Being There: Putting Brain, Body and World Together Again*. Cambridge, Mass.: MIT Press.
Clarke, Bruce. 2020. *Gaian Systems: Lynn Margulis, Neocybernetics, and the End of the Anthropocene*. Minneapolis: University of Minnesota Press.
Cruikshank, Julie. 2005. *Do Glaciers Listen? Local Knowledge, Colonial Encounters, and Social Imagination*. Vancouver: University of British Columbia Press.
Curtin, Deane. 1996. "Toward an Ecological Ethic of Care." In *Ecological Feminist Philosophies,* edited by Karen J. Warren, 66–81. Bloomington: Indiana University Press.
Danowski, Déborah, and Eduardo Viveiros de Castro. 2017. *The Ends of the World*. Cambridge: Polity.
Darwin, Charles. 1884. *The Origin of Species by Means of Natural Selection; or, The Preservation of Favoured Races in the Struggle for Life*. 6th ed. London: John Murray.
Derrida, Jacques. 1992. *Given Time: 1. Counterfeit Money*. Chicago: University of Chicago Press.
Derrida, Jacques. 2008. *The Animal That Therefore I Am*. New York: Fordham University Press.
Evernden, Neil. 1999. *The Natural Alien: Humankind and the Environment*. Toronto: University of Toronto Press.
Fergusson, David. 2018. *The Providence of God: A Polyphonic Approach*. Cambridge: Cambridge University Press.
Foltz, Bruce. 1995. *Inhabiting the Earth: Heidegger, Environmental Ethics and the Metaphysics of Nature*. London: Humanities Press.
Foucault, Michel. 1986. *The History of Sexuality, Vol. 3: The Care of the Self*. London: Allen Lane.
Gilligan, Carol. 1982. *In a Different Voice: Psychological Development and Women's Development*. Cambridge, Mass.: Harvard University Press.
Golinski, Jan. 2007. *British Weather and the Climate of Enlightenment*. Chicago: University of Chicago Press.
Gould, Stephen Jay. 2002. *The Structure of Evolutionary Theory*. Cambridge, Mass.: Harvard University Press.
Graham, Peter. 2020. *Traces of (Un)sustainability: Towards a Materially Engaged Ecology of Mind*. New York: Peter Lang.
Gray, John. 2004. "An Illusion with a Future." *Daedalus* 133 (3): 10–17.
Haar, Michel. 1993. *The Song of the Earth: Heidegger and the Grounds of the History of Being*. Bloomington: Indiana University Press.
Harding, Stephan. 2006. *Animate Earth: Science, Intuition and GAIA*. White River Junction, Vt.: Chelsea Green.

Heidegger, Martin. 1977a. "Letter on Humanism." In *Basic Writings,* 193–242. New York: Harper and Row.

Heidegger, Martin. 1977b. "The Question Concerning Technology." In *Basic Writings,* 283–318. San Francisco: Harper.

Heidegger, Martin. 1989. *Contributions to Philosophy (of the Event).* Indianapolis: Indiana University Press.

Heidegger, Martin. 1993. "Building, Dwelling, Thinking." In *Basic Writings,* edited by David Farrell Krell, 319–40. New York: HarperCollins.

Heidegger, Martin. 1995. *The Fundamental Concepts of Metaphysics: World, Finitude, Solitude.* Bloomington: Indiana University Press.

Heidegger, Martin. 1998. "On the Essence and Concept of φύσις." In *Pathmarks,* 183–230. Cambridge: Cambridge University Press.

Hochschild, Arlie Russell. 1983. *The Managed Heart: Commercialization of Human Feeling.* Berkeley: University of California Press.

Jefferies, Richard. 1881. *Wood Magic: A Fable.* London: Cassell, Petter, Galpin.

Jefferies, Richard. (1883) 1913. *The Story of My Heart: An Autobiography.* London: Longmans, Green.

Jefferies, Richard. 1934. *The Amateur Poacher.* London: Jonathan Cape.

Jefferies, Richard. 1939. *After London and Amaryllis at the Fair.* London: J. M. Dent.

Jefferies, Richard. 1947. *The Life of the Fields.* London: Lutterworth Press.

Jefferies, Richard. 1948. *The Notebooks of Richard Jefferies.* Edited by Samuel J. Looker. London: Grey Walls Press.

Kant, Immanuel. 1952. *Critique of Judgement.* Oxford: Oxford University Press.

Kant, Immanuel. 1987. *Critique of Judgement.* Indianapolis, Ind.: Hackett.

Kant, Immanuel. 1991. "Idea for a Universal History with a Cosmopolitan Purpose." In *Political Writings,* edited by Hans Reiss, 41–53. Cambridge: Cambridge University Press.

Kimmerer, Robin Wall. 2013. *Braiding Sweetgrass: Indigenous Wisdom, Scientific Knowledge and the Teachings of Plants.* Minneapolis, Minn.: Milkweed.

Klein, Naomi. 2007. *The Shock Doctrine: The Rise of Disaster Capitalism.* Toronto: Alfred A. Knopf.

Kolnai, Aural. 2004. *On Disgust.* Chicago: Open Court.

Larrabee, Mary Jeanne, ed. 1993. *An Ethic of Care: Feminist and Interdisciplinary Perspectives.* New York: Routledge.

Latour, Bruno. 1993. *We Have Never Been Modern.* Cambridge, Mass.: Harvard University Press.

Latour, Bruno. 1998. "To Modernize or Ecologise? That Is the Question." In *Remaking Reality: Nature at the Millennium,* edited by Bruce Braun and Noel Castree, 221–42. London: Routledge.

Latour, Bruno. 2017a. *Facing Gaia: Eight Lectures on the New Climatic Regime*. Oxford: Polity.
Latour, Bruno. 2017b. "Why Gaia Is Not a God of Totality." *Theory, Culture, and Society* 34 (2–3): 61–81.
Latour, Bruno. 2018. *Down to Earth: Politics in the New Climatic Regime*. Cambridge: Polity.
Latour, Bruno. 2020. "This Is a Global Catastrophe That Has Come from Within." *The Guardian,* June 6. https://www.theguardian.com/world/2020/jun/06/bruno-latour-coronavirus-gaia-hypothesis-climate-crisis.
Levinas, Emmanuel. 1991. *Totality and Infinity*. Dordrecht, Netherlands: Kluwer.
Liotta, P. H., and Allan W. Shearer. 2007. *Gaia's Revenge: Climate Change and Humanity's Loss*. Westport, Conn.: Praeger.
Lefebvre, Henri. 1994. *The Production of Space*. Oxford: Blackwell.
Lloyd, Genevieve. 2008. *Providence Lost*. Cambridge, Mass.: Harvard University Press.
Lovelock, James. 1979. *Gaia: A New Look at Life on Earth*. Oxford: Oxford University Press.
Lovelock, James. 2000. *Gaia: The Practical Science of Planetary Medicine*. Oxford: Oxford University Press.
Lovelock, James. 2006. *The Revenge of Gaia: Why the Earth Is Fighting Back and How We Can Still Save Humanity*. London: Penguin.
Lovelock, James. 2009. *The Vanishing Face of Gaia: A Final Warning*. London: Allen Lane.
Lovelock, James. 2019. *Novacene: The Coming Age of Hyperintelligence*. London: Allen Lane.
Macfarlane, Robert. 2019. *Underland: A Deep Time Journey*. London: Hamish Hamilton.
MacGregor, Sherilyn. 2006. *Beyond Mothering Earth: Ecological Citizenship and the Politics of Care*. Vancouver: University of British Columbia Press.
Malafouris, Lambros. 2013. *How Things Shape the Mind: A Theory of Material Engagement*. Cambridge, Mass.: MIT Press.
Manolopoulos, Mark. 2009. *If Creation Is a Gift*. Albany: State University of New York Press.
Margulis, Lynn. 1997. "Big Trouble in Biology: Physiological Autopoiesis versus Mechanistic Neo-Darwinism." In *Slanted Truth: Essays on Gaia, Symbiosis and Evolution,* edited by Lynn Margulis and Dorian Sagan, 265–82. NewYork: Springer.
Marion, Jean-Luc. 2002. *Being Given: Towards a Phenomenology of Giveness*. Stanford, Calif.: Stanford University Press.
Merchant, Caroline. 2006. *Earthcare: Women and the Environment*. London: Routledge.

Mill, John Stuart. 1885. *Three Essays on Religion: Nature, the Utility of Religion, and Theism*. London: Longmans, Green.

Molesky, Mark. 2016. *This Gulf of Fire: The Great Lisbon Earthquake; or, Apocalypse in the Age of Science and Reason*. New York: Vintage Books.

Morton, Timothy. 2007. *Ecology without Nature*. Cambridge, Mass.: Harvard University Press.

Nancy, Jean-Luc. 1991. *The Inoperative Community*. Minneapolis: University of Minnesota Press.

Nancy, Jean-Luc. 2007. *The Creation of the World or Globalization*. Albany: State University of New York Press.

Nitecki, Mathew H., ed. 1988. *Evolutionary Progress*. Chicago: University of Chicago Press.

Oslington, Paul. 2011. "Divine Action, Providence, and Adam Smith's Invisible Hand," in *Adam Smith as Theologian*, ed. Paul Oslington, 61–74. London: Routledge.

Plumwood, Val. 2013. *The Eye of the Crocodile*. Canberra: Australian National University Press.

Polt, Richard. 2007. "Ereignis." In *A Companion to Heidegger*, edited by Hubert L. Dreyfus and Mark A. Wrathall, 375–89. Oxford: Blackwell.

Povinelli, Elizabeth A. 1995. "Do Rocks Listen? The Cultural Politics of Apprehending Australian Aboriginal Labor." *American Anthropologist* 97 (3): 505–18.

Povinelli, Elizabeth A. 2016. *Geontologies: A Requiem to Late Liberalism*. Durham, N.C.: Duke University Press.

Ray, John. 1691. *The Wisdom of God as Manifested in the Works of the Creation*. London: Samuel Smith.

Richards, Robert J. 2002. *The Romantic Conception of Life: Science and Philosophy in the Age of Goethe*. Chicago: University of Chicago Press.

Roach, Catherine M. 2003. *Mother/Nature: Popular Culture and Environmental Ethics*. Bloomington: Indiana University Press.

Rudwick, Martin. 2014. *Earth's Deep History: How It Was Discovered and Why It Matters*. Chicago: University of Chicago Press.

Ruse, Michael. 1988. "Molecules to Men: Evolutionary Biology and Thoughts of Progress" in *Evolutionary Progress*, ed. Matthew H. Nitecki, 97–126. Chicago: University of Chicago Press.

Ruse, Michael. 2005. "Darwinism and Mechanism: Metaphor in Science." *Studies in the History and Philosophy of Biological and Biomedical Science* 36: 285–302.

Ruse, Michael. 2013. *The Gaia Hypothesis: Science on a Pagan Planet*. Chicago: University of Chicago Press.

Sallis, John. 2016. *The Figure of Nature: On Greek Origins*. Bloomington: Indiana University Press.

Sandilands, Catriona. 1999. *The Good-Natured Feminist: Ecofeminism and the Quest for Democracy*. Minneapolis: University of Minnesota Press.

Sands, Danielle. 2015. "Gaia, Gender and Sovereignty in the Anthropocene." *philoSOPHIA* 5 (2): 287–307.

Serres, Michel. 1995. *Genesis*. Ann Arbor: University of Michigan Press.

Shakespeare, William. 1953. "The Tragedy of Macbeth." In *The Works of Shakespeare*, vol. 3, 489–558. London: Nonesuch Press.

Smith, Mick. 2001. "Lost for Words? Gadamer and Benjamin on the Nature of Language and the 'Language' of Nature." *Environmental Values* 10 (1): 59–75.

Smith, Mick. 2005. "Hermeneutics and the Culture of Birds: The Environmental Allegory of 'Easter Island.'" *Ethics, Place, and Environment* 8 (1): 21–38.

Smith, Mick. 2010. "Epharmosis: Jean-Luc Nancy and the Political Oecology of Creation." *Environmental Ethics* 32 (4): 385–404.

Smith, Mick. 2011. "Dark Ecology." *Environmental Politics* 20 (1): 33–138.

Smith, Mick. 2012. "The Earthly Politics of Ethical An-arche: Arendt, Levinas, and Being with Others." In *Facing Nature: Levinas and Environmental Thought*, edited by William Edelglass, James Hatley, and Christian Diehm, 36–64. Pittsburgh, Pa.: Duquesne University Press.

Smith, Mick. 2013. "Ecological Community, the Sense of the World, and Senseless Extinction." *Environmental Humanities* 2: 21–41.

Smith, Mick. 2017. "Rain." In *Veer Ecology: An Ecotheory Companion*, edited by Jeffrey Cohen and Lowell Duckert, 230–45. Minneapolis: University of Minnesota Press.

Smith, Mick. 2019. "(A)wake for 'the Passions of This Earth': Extinction and the Absurd 'Ethics' of Novel Ecosystems." *Cultural Studies Review* 25 (1): 119–34.

Smith, Mick, and Joyce Davidson. 2006. "'It Makes My Skin Crawl . . .': The Embodiment of Disgust in Phobias of 'Nature.'" *Body and Society* 12 (1): 43–67.

Spretnak, Charlene. 1986. *The Spiritual Dimension of Green Politics*. Santa Fe, N.M.: Bear.

Starhawk. 1999. *Spiral Dance: A Rebirth of the Ancient Religion of the Great Goddess*. New York: HarperCollins.

Stengers, Isabelle. 2018. "The Challenge of Ontological Politics." In *A World of Many Worlds*, edited by Marisol De la Cadena and Mario Blaser, 83–111. Durham, N.C.: Duke University Press.

Sullivan, Sian. 2010. "'Ecosystem Service Commodities'—a New Imperial Ecology? Implications for Animist Immanent Ecologies, with Deleuze and Guattari." *New Formations* 69: 111–28.

Sutton, John. 2020. "Place and Memory: History, Cognition, Phenomenology." In *The Geography of Embodiment in Early Modern England*, edited by Garrett Sullivan and Mary Floyd-Wilson. Oxford: Oxford University Press.

Taylor, Charles. 2005. *Modern Social Imaginaries*. Durham, N.C.: Duke University Press.

Thomas, Edward. 1909. *Richard Jefferies: His Life and Work*. London: Hutchinson.

Thomson, Keith. 2005. *The Watch on the Heath: Science and Religion before Darwin*. London: HarperCollins.

Vogel, Lawrence. 2018. "Evolution and the Meaning of Being: Heidegger, Jonas and Nihilism." *Continental Philosophy Review* 51: 65–79.

Wain, John. 1974. *Samuel Johnson: A Biography*. New York: Viking Press.

Ward, Colin, and David Goodway. 2003. *Talking Anarchy*. Nottingham, U.K.: Five Leaves.

Weber, Max. 1964. *The Theory of Social and Economic Organization*. New York: Free Press.

Weiner, Norbert. 1948. *Cybernetics; or, Control and Communication in the Animal and the Machine*. Cambridge, Mass.: MIT Press.

Wilkins, John S. 2009. *Species: A History of the Idea*. Berkeley: University of California Press.

Wolfe, Cary. 1998. *Critical Environments: Postmodern Theory and the Pragmatics of the "Outside."* Minneapolis: University of Minnesota Press.

Wolfe, Cary. 2010. *What Is Posthumanism?* Minneapolis: University of Minnesota Press.

Young, Jason. 2020. "Bridging Bateson's Gap: Participating Cybernetically in a More-than-Human World." *Cybernetics and Human Knowing* 27 (2): 27–40.

Zimmerman, Michael E. 1990. *Heidegger's Confrontation with Modernity: Technology, Politics, Art*. Indianapolis: Indiana University Press.

Žižek, Slavoj. 2008. *In Defense of Lost Causes*. London: Verso.

Zuckert, Rachel. 2007. *Kant on Beauty and Biology: An Interpretation of the Critique of Judgement*. Cambridge: Cambridge University Press.

(Continued from page iii)

Forerunners: Ideas First

Claudia Milian
LatinX

Aaron Jaffe
Spoiler Alert: A Critical Guide

Don Ihde
Medical Technics

Jonathan Beecher Field
Town Hall Meetings and the Death of Deliberation

Jennifer Gabrys
How to Do Things with Sensors

Naa Oyo A. Kwate
Burgers in Blackface: Anti-Black Restaurants Then and Now

Arne De Boever
Against Aesthetic Exceptionalism

Steve Mentz
Break Up the Anthropocene

John Protevi
Edges of the State

Matthew J. Wolf-Meyer
Theory for the World to Come: Speculative Fiction and Apocalyptic Anthropology

Nicholas Tampio
Learning versus the Common Core

Kathryn Yusoff
A Billion Black Anthropocenes or None

Kenneth J. Saltman
The Swindle of Innovative Educational Finance

Ginger Nolan
The Neocolonialism of the Global Village

Joanna Zylinska
The End of Man: A Feminist Counterapocalypse

Robert Rosenberger
Callous Objects: Designs against the Homeless

William E. Connolly
Aspirational Fascism: The Struggle for Multifaceted Democracy under Trumpism

Chuck Rybak
UW Struggle: When a State Attacks Its University

Clare Birchall
Shareveillance: The Dangers of Openly Sharing and Covertly Collecting Data

la paperson
A Third University Is Possible

Kelly Oliver
Carceral Humanitarianism: Logics of Refugee Detention

P. David Marshall
The Celebrity Persona Pandemic

Davide Panagia
Ten Theses for an Aesthetics of Politics

David Golumbia
The Politics of Bitcoin: Software as Right-Wing Extremism

Sohail Daulatzai
Fifty Years of *The Battle of Algiers*: Past as Prologue

Gary Hall
The Uberfication of the University

Mark Jarzombek
Digital Stockholm Syndrome in the Post-Ontological Age

N. Adriana Knouf
How Noise Matters to Finance

Andrew Culp
Dark Deleuze

Akira Mizuta Lippit
Cinema without Reflection: Jacques Derrida's Echopoiesis and Narcissism Adrift

Sharon Sliwinski
Mandela's Dark Years: A Political Theory of Dreaming

Grant Farred
Martin Heidegger Saved My Life

Ian Bogost
The Geek's Chihuahua: Living with Apple

Shannon Mattern
Deep Mapping the Media City

Steven Shaviro
No Speed Limit: Three Essays on Accelerationism

Jussi Parikka
The Anthrobscene

Reinhold Martin
Mediators: Aesthetics, Politics, and the City

John Hartigan Jr.
Aesop's Anthropology: A Multispecies Approach

Mick Smith is professor in both philosophy and environmental studies at Queen's University, Ontario, and author of *Against Ecological Sovereignty: Ethics, Biopolitics, and Saving the Natural World* (Minnesota, 2011) and *An Ethics of Place: Radical Ecology, Postmodernity, and Social Theory.*

Jason Young is a PhD candidate in the School of Environmental Studies at Queen's University in Canada.